Metal–Insulator Transitions

Metal–Insulator Transitions

Second Edition

N. F. MOTT

Emeritus Cavendish Professor of Physics
University of Cambridge

Taylor & Francis
London • New York • Philadelphia

UK	Taylor & Francis Ltd, 4 John St, London WC1N 2ET
USA	Taylor & Francis Inc., 1900 Frost Road, Suite 101, Bristol, PA 19007

British Library Cataloguing in Publication Data

Mott, N. F. (Nevill Francis)
   Metal–insulator transitions.—2nd ed
   1. Metal–insulator transitions
   I. Title
   530′.4′1

   ISBN 0-85066-783-6

Mott, N. F. (Nevill Francis), Sir
   Metal–insulator transitions/N. F. Mott.—2nd ed.
     p. cm.
   Includes bibliographical references.
   ISBN 0-85066-783-6
   1. Metal–insulator transitions. I. Title.
QC176.8.E4M68   1990
530.4′1—dc20

                                  90-10916
                                  CIP

Preface to the Second Edition

The first edition of this book, published in 1974, described metal–insulator transitions due to disorder (the Anderson transition), the Mott–Hubbard transition resulting from intra-atomic interaction (the Hubbard U), as well as certain other types such as band-crossir.g transitions and those of Verwey as in Fe_3O_4. Since then, our understanding of the Anderson transition has been completely transformed by the scaling theory of Abrahams et al. (1979), experiments that supported its conclusions and by the theory of interactions between electrons initiated by Altshuler and Aronov (1979) for the metallic state and by Efros and Shklovskii (1975) for hopping conduction. The part of the book that deals with these transitions, both in impurity bands and other systems, has been completely rewritten. Our description of the Mott–Hubbard transition has changed less, except that the transition in doped silicon and germanium is now believed to be of Anderson type, even though the formula for the critical concentration derived by the present author in 1949 for a Mott transition is in satisfactory agreement with experiment for almost all doped semiconductors. Our motive for a new edition dealing with the transition in crystalline conductors, particularly the transitional-metal oxides, is primarily the renewed interest in these materials generated by the discovery of the high-temperature superconductors. Some of these, it is widely believed, are antiferromagnetic insulators doped sufficiently heavily to become metallic. We include a chapter putting this point of view, in the context of what we know in general about metal–insulator transitions. We also describe our new understanding of the phenomenon in liquids.

In the first edition we expressed our debt to many colleagues with whom we have discussed the subject matter of this book, particularly J. V. Acrivos, P. W. Anderson, K. F. Berggren, M. H. Cohen, B. R. Coles, M. Cyrot, E. A. Davis, J. Friedel, L. Friedman, J. Goodenough, G. Gruner, P. Hagenmuller, S. Maekawa, T. M. Rice, M. Pollak, J. C. Thompson, and R. M. White. We now add the names of C. Cyrot-Lackmann, J. P. Doumerc, P. P. Edwards, M. Kaveh, J. Morgan, M. Pepper, G. A. Thomas and many others. We are also particularly grateful to many authors who contributed articles on aspects of the subject to a *Festschrift* presented to the author on the occasion of his eightieth birthday, without which parts of the book could hardly have been written.

We depart from the usual notation in two ways. The transfer integral, usually denoted by t, is written I, following our first edition and Mott and Davis (1979). Also, instead of "transition metal", we write "transitional metal" (following J. Friedel), in order to avoid confusion with the former word as used in the title of this book.

Nevill Mott

Contents

Introduction

An early success of quantum mechanics was the explanation by Wilson (1931a, b) of the reason for the sharp distinction between metals and non-metals. In crystalline materials the energies of the electron states lie in bands; a non-metal is a material in which all bands are full or empty, while in a metal one or more bands are only partly full. This distinction has stood the test of time; the Fermi energy of a metal, separating occupied from unoccupied states, and the Fermi surface separating them in k-space are not only features of a simple model in which electrons do not interact with one another, but have proved to be physical quantities that can be measured. Any metal–insulator transition in a crystalline material, at any rate at zero temperature, must be a transition from a situation in which bands overlap to a situation when they do not. Band-crossing metal–insulator transitions, such as that of barium under pressure, are described in this book.

Interest in metal–insulator transitions arose in the following way. At a conference in Bristol it was pointed out by de Boer and Verwey (1937) that nickel oxide, a transparent non-metal, should, according to the Wilson model, be metallic because the eight electrons of the Ni^{2+} ion would only partly fill one of the d-bands. Peierls, in the discussion, said he thought that this must be due to correlation and gave a description of how this could occur. At the time we did not know that nickel oxide was antiferromagnetic; as Slater pointed out in 1951, an antiferromagnetic lattice can split the d-bands, allowing all bands to be full or empty. This description of antiferromagnetic insulators is discussed in Chapter 4 of this book, and is in our view legitimate at zero temperature; however, antiferromagnetic insulators normally retain their non-metallic properties above the Néel temperature, so a superlattice is not the whole story.

The present author (Mott 1949, 1956, 1961) described a metal–insulator transition by imagining a crystalline array of hydrogen-like atoms with a lattice constant a that could be varied. The example of nickel oxide suggested that for large values of a the material would be insulating, while the example of monovalent metals like sodium showed that for small values it would be metallic.

1

The question was, at what value of a would a metal–insulator transition occur? The assumption (Mott 1949) was made that this would occur when the screened potential round each positive charge,

$$V(r) = -\frac{e^2}{r} e^{-qr}, \tag{1}$$

with the screening constant q calculated by the Thomas–Fermi method, was just strong enough to trap an electron. The transition would be discontinuous; with varying n, and thus q, there would be a discontinuous transition from the state with all electrons trapped to that where all are free. It was found that this would occur when

$$n^{1/3} a_H = 0.2 \tag{2}$$

where n is the number of electrons per unit volume and a_H the hydrogen radius. Equation (2) was applied, with success, to the metal–insulator transition in doped semiconductors, the disorder resulting from the random positions of the centres being neglected. The disorder was believed to be the cause of the absence of a discontinuous change in the conductivity. However, the success of (2) was fortuitous; if the many-valley nature of the conduction bands of silicon and germanium is taken into account, and also the central cell corrections, totally different results are obtained (Kreiger and Nightingale 1971, Martino 1973, Green et al. 1977).

The next important step in the development of the theory was the introduction by Hubbard (1964a, b) of a model in which the interaction between electrons is included only when they are on the same atom, long-range Coulomb forces being neglected. For large values of a this model leads to an antiferromagnetic insulator; this splits the band, so that, much as in Slater (1951), an electron's energies lie in a full band and an empty one. However, Hubbard's work was criticised by Herring (1966) for lack of a clear recognition that the low-density form is antiferromagnetic. More recently, work on "resonating valence bond" states has shown that an antiferromagnetic lattice is not essential in the insulating state (see Section 3.3).

When the two bands exist, we call them the lower and upper Hubbard bands. The metal–insulator transition then appears as a band-crossing transition, from an antiferromagnetic metal to an antiferromagnetic insulator. There is another transition from the antiferromagnetic metal to the normal metal. This "normal" metal, however, is highly correlated, and this leads to a great enhancement of the electronic specific heat and Pauli susceptibility, as first pointed out by Brinkman and Rice (1970b) and observed in vanadium oxides. The Hubbard model, which does not include long-range forces, does not predict a discontinuous change in n, the number of carriers. When long-range forces are included, as first shown by Brinkman and Rice (1973), for a crystalline system a discontinuity will always occur when a full and empty band cross, as a result of the long-range electron–hole interaction, whether the bands are of Hubbard type or those generated by the crystal structure. Mott (1978) has discussed the conditions under which disorder can remove this discontinuity. Actually, however, in many-valley doped semiconductors, the transition now appears to be of what we now call Anderson type and not directly a consequence of short-range (intra-atomic) interaction, both disorder and long-range interaction contributing to an equation similar to (2).

For disordered systems, then, a quite different form of metal–insulator transition occurs—the Anderson transition. In these systems a range of energies exists in which the electron states are localized, and if at zero temperature the Fermi energy lies in this range then the material will not conduct, even though the density of states is not zero. The Anderson transition can be discussed in terms of non-interacting electrons, though in real systems electron–electron interaction plays an important part.

The plan of this book is as follows. Chapter 1, after some introductory sections on band theory and conductivity, deals with the phenomena that can be described in terms of a model of non-interacting electrons, including metal–insulator transitions due to band crossing such as that shown by some divalent metals under pressure. It also describes the behaviour of electrons in a non-crystalline medium and the Anderson transition, including that in liquids, in a model where interactions are neglected. Chapter 2 discusses first the interaction of electrons with phonons, particularly the mass enhancement by polaron formation. Then various phenomena due to electron–electron interaction are introduced, including exciton formation, the properties of excitons in metals, and the Nozières peak at an absorption edge, electron–electron collisions and their effect on the resistance (Landau–Baber scattering), and an introduction to the Hubbard intra-atomic energy U. Chapter 3 deals with magnetic moments; ferro- and antiferromagnetic coupling in non-metallic oxides are first described, followed by the Anderson–Wolff condition for the formation of moments in metals. The Kondo effect is relevant to some of the properties of materials near the metal–insulator transition, and an outline of the physics of this phenomenon is given. We also describe the interaction between moments in spin glasses and amorphous antiferromagnets. A brief section on metallic ferromagnets is introduced, mainly in order to discuss nearly ferromagnetic materials where the Pauli paramagnetism can be indefinitely enhanced without any large effect on the density of states or electronic specific heat; in this they are in strong contrast with the highly correlated nearly antiferromagnetic materials, where both are enhanced. Chapter 4 introduces the metal–insulator transition due to correlation, and starts with the Hubbard model in which only intra-atomic correlation is included. The sequence of antiferromagnetic insulator, antiferromagnetic metal and normal metal is described, and the comparison between the highly correlated nearly antiferromagnetic and nearly ferromagnetic metals is again made. In the former the spin-flip on singly occupied centres is compared with the Kondo behaviour.

In Chapter 5 we return to the Anderson transition resulting from disorder, and describe the work done in the last decade on the major effects caused by electron–electron long-range interaction. These we compare with experimental work on the electrical and other properties of electrons in impurity bands in doped semiconductors. Chapter 6 considers the wide variety of transitions that can occur in transitional-metal oxides: band-crossing, Mott, Hubbard and Verwey. Chapter 7 reviews such materials as tungsten bronzes, vanadium monoxide and $La_{1-x}Sr_xVO_3$, where disorder probably plays a role. Chapter 8 deals with Wigner and Verwey transitions, and Chapter 9 contains a tentative application of these ideas to high-temperature superconductors. Chapter 10 deals with the transition in liquids, including the somewhat controversial problem of whether quantum interference occurs in liquid metals.

In this book we confine ourselves to three-dimensional systems. According to scaling theory, in two-dimensional systems all states are localized. Although it has been suggested that a transition between power-law and exponential localization can occur (Kaveh 1985c, Mott and Kaveh 1985a, b), this matter is too uncertain at present for discussion in a book.

Throughout this book, particularly in the later chapters, we assume that a condensed electron gas can be treated as a Fermi liquid of pseudoparticles, for instance dielectric or spin polarons. We recognize that this is an unproved assumption.

1

Non-Interacting Electrons

1 Free electrons in metals

The first application of quantum mechanics to electrons in solids is contained in a paper by Sommerfeld published in 1928. In this the free-electron model of a metal was introduced, and, for so simple a model, it was outstandingly successful. The assumptions made were the following. All the valence electrons were supposed to be free, so that the model neglected both the interaction of the electrons with the atoms of the lattice and with one another, which is the main subject matter of this book. Therefore each electron could be described by a wave function ψ identical with that of an electron in free space, namely

$$\psi = A e^{i\mathbf{k} \cdot \mathbf{r}},$$

where, if the volume of the metal is Ω,

$$A = \Omega^{-1/2}.$$

The vector $\mathbf{k}$ is here the wave vector describing the momentum of the electron. But, unlike those for electrons in free space, the values that $\mathbf{k}$ can have are quantized; in a cube of side L ($\Omega = L^3$), if we write

$$\mathbf{k} \cdot \mathbf{r} = k_1 x + k_2 y + k_3 z$$

then k_1, k_2, k_3 can have the values

$$k_1 = \frac{2\pi l_1}{L}, \quad k_2 = \frac{2\pi l_2}{L}, \quad k_3 = \frac{2\pi l_3}{L}, \tag{1}$$

where l_1, l_2, l_3 are positive or negative integers. Starting from this concept of quantization, one can introduce the "density of states". In k-space the number of states in a volume element d^3k ($= dk_x dk_y dk_z$) is

$$\Omega \, d^3k / 8\pi^3,$$

5

and the number of states for which k, the modulus of $\mathbf{k}$, lies in the range k to $k+dk$ is

$$4\pi\Omega k^2 \, dk/8\pi^3. \tag{2}$$

The density of states per unit energy range and per unit volume, for given spin direction, is written $N(E)$, where E denotes the energy. Thus from (2), setting $\Omega = 1 \text{ cm}^3$, we can write

$$N(E)\,dE = 4\pi k^2 \, dk/8\pi^3,$$

and since the relation between E and k is

$$E = \hbar^2 k^2/2m,$$

we find, substituting for k,

$$N(E) = \frac{1}{4\pi^2}\left(\frac{2m}{\hbar^2}\right)^{3/2} E^{1/2}. \tag{3}$$

The successes of the free-electron model came from combining it with Fermi–Dirac statistics, according to which the number of electrons in each orbital state cannot be greater than two, one for each spin direction. Thus at the absolute zero of temperature all states are occupied up to a maximum energy E_F given by

$$2\int^{E_F} N(E)\,dE = n,$$

where n is the number of electrons per unit volume. This gives, on making use of (3),

$$\left(\frac{2mE_F}{\hbar^2}\right)^{3/2}\bigg/3\pi^2 = n. \tag{4}$$

E_F is called the *Fermi energy*, and in most metals its magnitude is several electron volts. The sphere in k-space separating occupied from unoccupied states is called the *Fermi surface*, and its radius k_F is given by

$$\frac{8\pi}{3}k_F^3/8\pi^3 = n, \tag{5}$$

whence approximately

$$k_F \approx \pi n^{1/3}.$$

At a finite temperature T the number of electrons with energies between E and $E+dE$ is

$$2N(E)f(E)\,dE,$$

where $f(E)$ is the Fermi–Dirac distribution function defined by

$$f(E)=\left[\exp\left(\frac{E-\zeta}{k_{\mathrm{B}}T}\right)+1\right]^{-1}$$

(k_{B} is Boltzmann's constant). Here ζ, the chemical potential of the gas, tends to E_{F} as T tends to zero.

A major achievement of the free-electron model was to show why the contributions of the free electrons to the heat capacity and magnetic susceptibility of a metal are so small. According to Boltzmann statistics, the contribution to the former should be $\frac{3}{2}nk_{\mathrm{B}}$ per unit volume. According to Fermi–Dirac statistics, on the other hand, only a fraction of order $k_{\mathrm{B}}T/E_{\mathrm{F}}$ of the electrons acquire any extra energy at temperature T, and these have extra energy of order $k_{\mathrm{B}}T$. Thus the specific heat is of order $nk_{\mathrm{B}}^{2}T/E_{\mathrm{F}}$, and an evaluation of the constant gives

$$c_{v}=\tfrac{1}{2}\pi^{2}nk_{\mathrm{B}}^{2}T/E_{\mathrm{F}},$$

and, since $k_{\mathrm{B}}T\ll E_{\mathrm{F}}$, this is small. In the same way, each excited electron contributes $\sim\mu^{2}/k_{\mathrm{B}}T$ to the susceptibility, which can be shown to be

$$\tfrac{3}{2}n\mu^{2}/E_{\mathrm{F}}.$$

This quantity is called the Pauli spin susceptibility.

2 Electrons in the conduction band of a crystal

2.1 Bloch's theorem and Brillouin zones

The solution of the Schrödinger equation

$$\nabla^{2}\psi+\frac{2m}{\hbar^{2}}[E-V(x,y,z)]\psi=0 \tag{6}$$

with a periodic potential energy $V(x,y,z)$ has been the subject of a very large number of original papers and reviews. Here we shall give only an outline of some outstanding results relevant to our main theme. The problem is divisible in two parts:

(a) an estimation of the best potential, for instance the Hartree–Fock potential;

(b) the calculation of the allowed energies E and of the wave functions ψ.

This section will deal mainly with the second aspect of the problem.

All considerations depend on the theorem of Bloch (1929), according to which all solutions of (6) are of the form

$$\psi_{k}=\mathrm{e}^{\mathrm{i}k\cdot r}u_{k}(x,y,z), \tag{7}$$

where $u_{k}(x,y,z)$ has the same period as the lattice. The corresponding energy will be written $E(\mathbf{k})$. Each wave function is thus characterized by a wave vector $\mathbf{k}$ and represents an electron moving through the lattice without being scattered. The fact that each electron is described by a solution of this kind shows why pure metals at low temperatures have high conductivity; resistivity arises only from scattering by impurities or defects in the lattice, by phonons and by other electrons (which normally gives a small contribution as shown in Chapter 2, Section 6).

The concept of k-space is thus useful here, too, each value of $\mathbf{k}$ satisfying the quantum condition (1) defining a state of the electron. For a given crystal structure we can define in k-space a polyhedron called the "Brillouin zone"; in cubic structures the Brillouin zone contains just one electron per atom for each spin direction, or two electrons per atom including both spin directions. We define the wave vector $\mathbf{k}$ only within this zone, so that the energies should be written $E_n(\mathbf{k})$, n being 1 for the first zone, 2 for the second zone and so on. Within a given zone, and thus for a given value of n, $E_n(\mathbf{k})$ is a continuous function of $\mathbf{k}$; also, except at points of particular symmetry where degeneracies may exist, $E_n(\mathbf{k})$ is never equal to $E_{n'}(\mathbf{k})$.

The relation of the Brillouin zone to the crystal structure is discussed in many textbooks on solid-state physics (see e.g. Jones 1960, Ziman 1964, p. 12, Ashcroft and Mermin 1976, Chap. 8).

2.2 Methods of calculating energy bands

2.2.1 The tight-binding approximation; s-bands

One of the first ways used to calculate approximate solutions of (6) was the "tight-binding" method. Here we consider an array of N potential wells, of which two are shown in Fig. 1.1, and suppose that in each well individually the electrons can have a number of bound states with energies $W_0, W_1, \ldots, W_n$ and with wave functions $\phi_0, \phi_1, \ldots, \phi_n$. When the electron is allowed to move from one well to another, a band containing N states is formed from each bound state of a single well. The approximate wave function describing this motion is

$$\psi_{nk} = N^{-1/2} \sum_i e^{i\mathbf{k}\cdot\mathbf{a}_i} \phi_{ni}(\mathbf{r} - \mathbf{a}_i). \tag{8}$$

Here $\mathbf{a}_i$ denotes the lattice point at which a well is situated. The value $E_n(\mathbf{k})$ of the energy of an electron with this wave function is

$$E_n(\mathbf{k}) = W_n + \int \psi_{nk}^* \Delta V \psi_{nk} \, \mathrm{d}^3 x,$$

where ΔV is the difference between the potential energy V and that of a single well (Fig. 1.1). If n refers to an s-state and the lattice structure is simple cubic then the energy of the state $\mathbf{k}$ is

$$E_n(\mathbf{k}) = W_n - \beta - 2I(\cos k_x a + \cos k_y a + \cos k_z a). \tag{9}$$

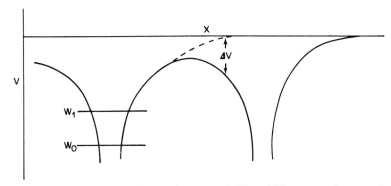

Fig. 1.1 Potential energy of an electron in a crystal. W_0 and W_1 are atomic energy levels.

Here

$$\beta = -\int |\phi_i|^2 \, \Delta V \mathrm{d}^3 x,$$

and I is the "overlap energy integral" defined by

$$I = -\int \phi_{i+1}^* \, \Delta V \phi_i \mathrm{d}^3 x. \tag{10}$$

Formulae for body- and face-centred structures are given in textbooks. A general result is that the bandwidth B is given by

$$B = 2zI,$$

where z is the coordination number.

For 1s-states, for which $\phi = (\alpha^3/\pi)^{1/2} \mathrm{e}^{-\alpha r}$, I has been evaluated (Slater 1963) and gives

$$I = \frac{e^2 \alpha}{\kappa} (1 + \alpha R) \mathrm{e}^{-\alpha R}, \tag{11}$$

where R is the distance between the centres, and the potential energy of an electron in one well is $-e^2/\kappa r$.

In deriving these formulae it is assumed that the wave functions ϕ_i are orthogonal to one another, as for instance are Wannier functions. If they are atomic functions falling off as $\mathrm{e}^{-\alpha r}$, a formula such as (9) must be modified by a term in the denominator to take account of the non-orthogonality. Reviews of the appropriate formulae are given in textbooks; see e.g. Callaway (1964) and Wohlfarth (1953), who considered a linear chain of hydrogen atoms.

One application of these formulae in this book is to impurity bands in doped silicon or germanium. Here the centres are distributed at random; the appropriate formulae for this case are discussed in Section 7 and Chapter 6.

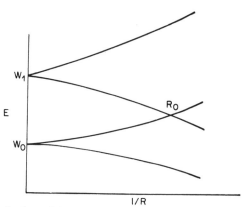

Fig. 1.2   Allowed values of the energy of an electron in a crystal, plotted against $1/R$ where R is the distance between atoms. In a theory of non-interacting electrons a metal–insulator transition takes place when $R = R_0$.

Figure 1.2 represents schematically the way in which two atomic levels broaden into bands as the distance between the atoms decreases. In three or two dimensions the bands can overlap, as shown in the figure, though this is not possible in one dimension.

A natural extension of the tight-binding approximation is to write

$$\Psi = \sum_n A_{nk}\psi_{nk}, \tag{12}$$

where the ψ_{nk} are the functions already defined by (8). The method is then identical with the "linear combination of atomic orbitals" (LCAO) used in theoretical chemistry. For energy bands in metals, apart from d-bands, the method converges slowly and is of little practical use; however, for d-bands, whether in metals or oxides, it is convenient. These will be described in Section 2.2.2. We first give a description, due to Mott and Jones (1936, pp. 72–76), of the treatment of s- and p-bands in a simple cubic lattice, which is important for our later treatment of some metal–insulator transitions.

We plot the energy E along the line $k_y = k_z = 0$ in the first and second Brillouin zones. The result is shown in Fig. 1.3, where the dotted lines show E as a function of k according to (9), and the full lines the result of hybridizing ψ_{1k} and ψ_{2k} according to (12). In ψ_{1k} the atomic orbitals ϕ could be 2s-functions and in ψ_{2k} they could be 2p-functions of the form $xf(r)$. Figure 1.4(a) shows what happens if the highest s-state lies below the lowest p-state. The top of the lower band is then wholly s-like; this is illustrated in Fig. 1.4. This will always be the case if the distance between the atoms is large, so that the overlap integrals are small. However, as the atoms come nearer together, the energy of the lowest Bloch function made up of p-states with principal quantum number n will in general drop below that of the most energetic function made up of s-states with the same principal quantum number; the hybridized function in the lowest band then goes continually from an s-like to a p-like form.

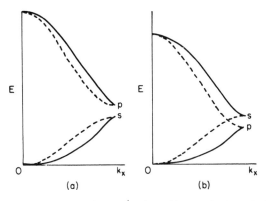

Fig. 1.3 Energy in s- and p-bands for a simple cubic metal plotted along the k_x axis; dashed lines without hybridization, full lines with hybridization: (a) atoms far apart; (b) atoms nearer together, so that (c) of Fig. 1.4 has lower energy than (b). From Mott and Jones (1936).

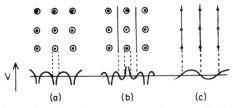

Fig. 1.4 Nodes and wave functions ψ of an electron in a metal: (a) at bottom of band; (b), (c) at $k_x = \pi/a$. (b) is s-like, (c) is p-like. From Mott and Jones (1936).

Metals of both types exist. That shown in Fig. 1.4(b) is probably the most common in divalent metals, but mercury (Mott 1972c) belongs to type (a). These conclusions will be discussed in terms of pseudopotentials in Section 2.2.4.

2.2.2 *The tight-binding approximation; d-bands*

The tight-binding approximation is useful for the calculation of d-bands, in both metals and oxides. As is well known, the three d-states with symmetry yz, zx, xy (called t_{2g} states) have energies that are split from those of the two states $x^2 - y^2$, $y^2 - z^2$ (called e_g) by a cubic crystal field. The question then is whether the splitting of the energy of the atomic d-states by the crystal field is greater than the overlap integral I. In transitional-metal oxides this is frequently the case. If so, we can set up bands derived from the t_{2g} functions and bands from the e_g functions. Both sets of bands are degenerate at $k = 0$, so the plot of energy E against k for the five bands is as in Fig. 1.5. Plots of this kind were first given by Jones and Mott (1937).

In metals, however, the overlap energy integral I for d-states is usually larger than the crystal-field splitting. Typical behaviour is shown in Fig. 1.6, which

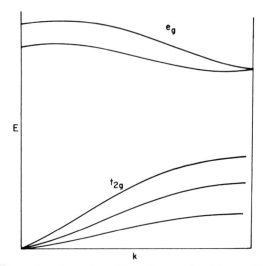

Fig. 1.5   Plot of energies of e_g and t_{2g} d-functions against k in a cubic lattice, for the case where the crystal-field splitting is large compared with the overlap energy integral.

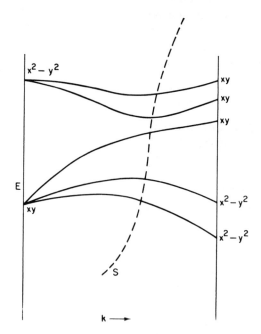

Fig. 1.6   Plot against k of energies of five d-bands when the bandwidths are greater than the crystal-field splitting, so that hybridization takes place between the xy and $x^2 - y^2$ functions. From Mott (1964).

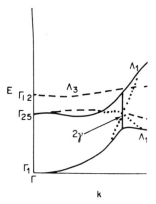

Fig. 1.7 Energy E plotted against k in the (111) direction for copper (schematic). The vertical line shows the value of k for which the s-band, calculated as for nearly free electrons, would cross the d-band (Heine 1969).

shows the t_{2g} and e_g bands after hybridization. The figure, taken from Mott (1964), is based on calculations by Asdente and Friedel (1961, 1962) for the body-centred cubic structure. The model clearly leads to a minimum in the density of states near the middle of the band, which is a marked feature of body-centred cubic alloys of transition metals. Hybridization with the s-band will also occur.

In transitional and noble metals the s-band crosses the d-band and is hybridized with it. The situation is discussed in a number of papers (cf. Mott 1964, Heine 1969, p. 25). Figure 1.7, taken from Heine (1969), shows the band structure of copper in the (111) direction. 2γ is the "hybridization gap" where the 4s-band crosses the d-bands.

2.2.3   Cellular and other methods

The first attempt to calculate realistic wave functions for electrons in metals is that of Wigner and Seitz (1933). These authors pointed out that space in a body- or face-centred cubic crystal could be divided into polyhedra surrounding each atom, that these polyhedra could be replaced without large error by spheres of radius r_0, so that for the lowest state one has to find spherically symmetrical solutions of the Schrödinger equation (6) subject to the boundary condition that

$$\partial\psi/\partial r = 0$$

on the boundary ($r = r_0$). Using this method, Wigner and Seitz obtained the well-known curve of E as a function of r_0 shown in Fig. 1.8.

Subsequent cellular methods, on which there is an enormous literature, will not be described here. We shall, however, need to introduce certain ideas, particularly that of the pseudopotential. We begin by introducing the concept of the muffin-tin potential due to Ziman (1964a). This is illustrated in Fig. 1.9. The tight-binding approximation is appropriate for states with energies below the muffin-tin zero ("bound bands" in Ziman's notation). If the energy is above the

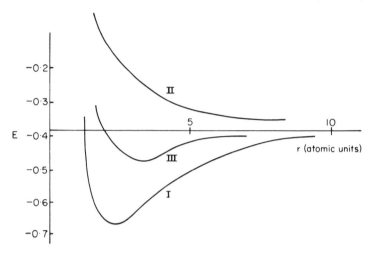

Fig. 1.8 Plot of energy E in metallic silver against atomic radius r: (I) energy E_0 of lowest state; (II) mean Fermi energy $\frac{3}{5}E_F$ of electron gas; (III) mean energy $E_0 + E_F$ of electron at the Fermi level. Energy in rydbergs (13.52 eV). From Mott and Jones (1936).

Fig. 1.9 The "muffin-tin" potential for an electron in a solid. The potential differs appreciably from zero in the shaded region. From Ziman (1964b).

muffin-tin lid then the wave function between the shaded regions can be built up of plane waves or alternatively of spherical harmonics. The whole band structure, then, will depend on the phase shifts η_l, functions of the energy, which are related to the logarithmic derivatives of the radial wave functions at $r = r_0$; the formula for the relationship is given by Ziman (1964a, p. 89). Following Heine (1969, p. 5), we write

$$\eta_l = p_l \pi + \delta_l, \tag{13}$$

where the integer p_l, chosen so that $|\delta_l| < 1$, is the number of radial nodes in the atomic wave function. If the phase shifts δ_l are all small then the wave functions behave as if they were nearly free-electron-like, but not otherwise.

It should be emphasized that in metals the d-states, for which tight-binding functions may be used, lie above the zero of the muffin-tin potential. The reason why the tight-binding method can still be used is the following. The radial part of the Schrödinger equation is

$$\frac{d^2\psi}{dr^2} + \frac{2}{r}\frac{d\psi}{dr} + \frac{2m}{\hbar^2}\left[E - V - \frac{l(l+1)}{r^2}\right]\psi = 0 \tag{14}$$

with $l = 2$. The term $l(l+1)/r^2$ acts as a repulsive potential round each atom, producing a potential barrier within the shaded area of Fig. 1.9.

2.2.4 The nearly-free-electron model; pseudopotentials

An early method of describing electrons in crystals was the method of "nearly free electrons"; we shall refer to it as the NFE model. In this the potential energy $V(x, y, z)$ in (6) is treated as small compared with the electron's total energy E. This is, of course, never the case in real crystals; the potential energy near the atomic core is always large enough to produce major deviations from the free-electron form. Therefore, until the introduction of the concept of a "pseudopotential", it was thought that the NFE model was not relevant to real crystalline solids.

The method is described in all textbooks on the subject. The essential point is the large deviations from the free-electron form

$$\psi = \text{const} \times e^{ikx}, \quad E = \frac{\hbar^2 k^2}{2m}$$

occur only when k is near those values for which Bragg reflection by the lattice can occur. The principles can be illustrated for a one-dimensional lattice of parameter a. By Bloch's theorem (7), we can write ψ in the form

$$\psi = e^{ikx}\sum_n A_n e^{-2\pi inx/a}. \tag{15}$$

In the Schrödinger equation (6) we take our zero of energy so that $\int V(x)\,dx = 0$. Inserting (15) into this Schrödinger equation, and *assuming* that all A_n are small compared with A_0, we find by the usual methods of perturbation theory

$$A_n = \frac{2mV_n}{\hbar^2}\frac{A_0}{k^2 - k_n^2}, \tag{16}$$

where

$$V_n = \int_0^a V(x)e^{2\pi inx/a}\frac{dx}{a}$$

and

$$k_n = k - 2\pi n/a.$$

Second-order perturbation theory gives

$$E = E_0 + \sum_{n \neq 0} \frac{|V_n|^2}{E_0 - E_n}, \tag{17}$$

where

$$E_0 = \frac{\hbar^2 k^2}{2m}, \quad E_n = \frac{\hbar^2 k_n^2}{2m}.$$

Equations (16) and (17) show that A_n/A_0 is not small when k and k_n are nearly equal, so that the assumption used is no longer valid. We then set

$$\psi = A_0 e^{ikx} + A_n e^{ik_n x}, \tag{18}$$

thereby neglecting all the A_n that are still small. Inserting (18) into the Schrödinger equation (6), multiplying by e^{-ikx} or $e^{ik_n x}$ and integrating from 0 to a, we find

$$A_0(E - E_0) - A_n V_n = 0,$$

$$-A_0 V_n + A_n(E - E_n) = 0.$$

Eliminating A_0, A_n from these equations, we obtain

$$(E - E_0)(E - E_n) - V_n V_n^* = 0,$$

which is a quadratic equation with the solutions

$$E = \tfrac{1}{2}\{E_0 + E_n \pm [(E_0 - E_n)^2 + 4V_n V_n^*]^{1/2}\}. \tag{19}$$

If we use (19) to plot E against k as in Fig. 1.10, a discontinuity ΔE given by

$$\Delta E = 2|V_n| \tag{20}$$

occurs. This is the band gap in one dimension. The extension to three dimensions is given in textbooks.

The wave functions at the points P and Q in Fig. 1.10 are standing waves of the form

$$\psi = \cos kx$$

and

$$\psi = \sin kx.$$

It will easily be seen that $\psi = \cos kx$ at P and $\sin kx$ at Q if V_n is negative. If V_n is positive, the reverse is the case. This conclusion should be compared with that obtained in Section 2.2.1 in our discussion of hybridized tight-binding wave functions; it is important for some metal–insulator transitions.

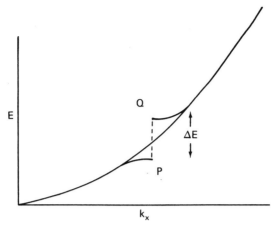

Fig. 1.10 Plot of energy against k_x in the approximation of nearly free electrons; ΔE is the energy gap.

The NFE model was brought back into the centre of band-theory calculations by the introduction at the beginning of the decade 1960–70 of the concept of the pseudopotential (for reviews see Ziman 1964b, Harrison 1966, Heine 1969, 1970). Following Heine (1969, p. 12), we define the pseudopotential $V(r)$ of each atom in a metal or semiconductor as the potential that will produce the phase shifts defined by (13). The "cancellation theorem" first discussed by Phillips and Kleinman (1959) showed why for most metals the radial nodes (Fig. 1.4) and the attractive potential give opposite and nearly equal effects, leading to small values of δ_l, so that $V(r)$ can be used to obtain band structures and other quantities (e.g. resistivity of liquid metals) using perturbation theory, as in the NFE approximation. Abarenkov and Heine (1965) introduced a form of the potential that is easy to visualize (Fig. 1.11), in which $V(r)$ is taken to be the Coulomb potential $-ze^2/r$ for large r, with a square-well repulsive potential replacing the provision that the wave function should have p_l nodes. The depth and radius of the square well are determined to give the best possible values of δ_l, and to orthogonalize the wave function with those of lower quantum number.

In practice the Fourier transform $v(q)$ of $V(r)$ (Heine 1969, p. 34) is the quantity that correlates most closely with observation. Some typical forms are shown in Fig. 1.12. Ziman (1964a, pp. 130, 177) has shown that $v(0)$ should be equal to $-\frac{2}{3}E_F$. According to (19), $2v(2k_0)$ is equal to the band gap at a given boundary of the Brillouin zone, k_0 being the value of k at this point in k-space.

For most metals $v(k_0)$ is *positive* at this point. This means, according to the considerations of Section 2.2.1, that the wave function is p-like at the boundary of the first zone, s-like at the bottom of the second zone, as illustrated in Figs. 1.3 and 1.4. Mercury, according to Evans (1970), is an exception. The reason proposed is that if d-states exist with energies near the Fermi level with a different principal quantum number then they hybridize with s-like states and lower their energies. In mercury these states are below the Fermi energy.

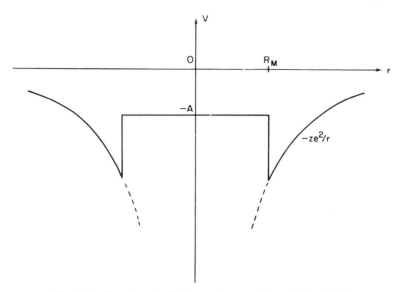

Fig. 1.11   Pseudopotential in real space. From Heine (1969).

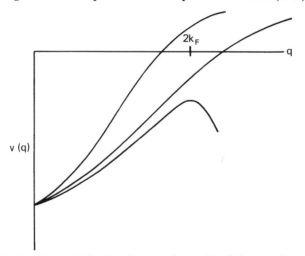

Fig. 1.12   Typical forms of the Fourier transform $v(q)$ of the pseudopotential plotted against q, the lower ones for larger lattice parameter.

For most recent developments in the use of pseudopotentials, useful papers are Hamann *et al.* (1979) and Cohen (1982). Cottrell (1988) has given a very readable outline of the present position.

For the theme of this book, namely the metal–insulator transition, it is important to realize that if the lattice parameter becomes larger, $v(q)$ will get progressively more negative, as shown in Fig. 1.12. Expansion of most metals will at first decrease the gap and then enlarge it; in mercury it will increase it from the beginning.

3 Metals and non-metals in the approximation of nearly free electrons

Wilson (1931a, b) introduced the model that predicted at the absolute zero of temperature a sharp distinction between crystalline materials showing metallic conduction, on the one hand, and insulating behaviour, on the other. Insulators according to this model are materials where all energy bands are completely occupied or completely empty. Since, for every occupied state with a wave function ψ of the form $e^{ikx}u(x, y, z)$, there is also an occupied state for which $\psi = e^{-ikx}u^*(x, y, z)$, no current can flow. A metal, on the other hand, is a material in which one or more energy bands are partly full. In such materials at zero temperature, states are full up to a limiting energy E_F (the Fermi energy) and states with higher energies are empty. For a perfect crystal the conductivity should tend to infinity as the temperature tends to zero. The density of states for a metal with one electron per atom, two electrons per atom and for an insulator is shown schematically in Fig. 1.13. For an insulator ΔW is the energy gap between the full band (known as the valence band) and the empty band (the conduction band); at a finite temperature the number of electrons and holes (current carriers) will be proportional to the factor $\exp(-\frac{1}{2}\Delta W/k_B T)$. We note that $N(E_F)$ must vanish for a crystalline insulator.

We believe that this formulation remains valid for crystalline materials at absolute zero, even if the materials are antiferromagnetic (like NiO), *provided* that the potential-energy function that is used in the Schrödinger equation (6) depends on the orientation of the spin of the electron considered, which determines the moment on a given atom. We can only make this statement at absolute zero; the fact that NiO, for instance, remains non-metallic above the Néel temperature requires separate consideration (Chapter 6, Section 2). It is well known, as we shall see in later chapters, that the application of band theory to "non-magnetic" NiO predicts metallic behaviour. At absolute zero, however, a crystal like NiO must have some magnetic or other superlattice with the symmetry to form bands containing a number of states, allowing them to be all either full or empty. But often it will not be sufficient, if we are to obtain even qualitatively the right separation between occupied and empty bands, to take the *same* potential $V(x, y, z)$ for the valence and conduction bands. Particularly in a tight-binding situation, as for a d-band, it would be a poor approximation to do so. An electron in the valence band of (say) solid argon sees the field of Ar^+, while an electron in the conduction band sees that of Ar. This should introduce a separation between the two bands of order $\mathscr{I}-\mathscr{E}$, where $\mathscr{I}$ is the ionization potential and $\mathscr{E}$ the electron affinity. This is a term depending on correlation, and if a one-electron formulation with the same function $V(r)$ for electrons in both bands gives good agreement with experiment for the energy gap then this appears as something of an accident.

In situations where $V(x, y, z)$ is not periodic, as for instance in the impurity band of doped semiconductors and in non-crystalline materials, it is still true that if $N(E_F)$ vanishes then the material is an insulator at zero temperature, but the converse is not true. This is because a finite value of $N(E_F)$, still within the context

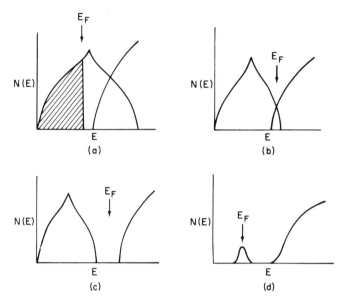

Fig. 1.13   Electron density of states $N(E)$ in a cubic material; E_F denotes the Fermi energy; (a) normal metal; (b) semimetal; (c) insulator; (d) n-type degenerate semiconductor.

of the theory of non-interacting electrons, is compatible with a situation in which all states are "localized" in the sense introduced by Anderson (1958). Under such conditions, $\sigma(\omega)$, the conductivity at frequency ω, is always finite at $T=0$ as long as ω does not vanish, but tends to zero roughly as ω^2 as $\omega \to 0$. The d.c. conductivity vanishes at zero temperature, and at finite but small T behaves as $\exp(-A/T^\nu)$, with ν between $\frac{1}{4}$ and $\frac{1}{2}$. This behaviour is discussed in Section 15.

Metal–insulator transitions in both crystalline and non-crystalline materials are often associated with the existence of magnetic moments. Moments on atoms in a solid are of course an effect of correlation, that is of interaction between electrons, and their full discussion is deferred until Chapter 3. But even within the approximation of non-interacting electrons in crystalline solids, metal–insulator transitions can occur. These will now be discussed.

4   Metal–insulator transitions caused by overlapping bands

A transition of this kind from metal to insulator will occur when some parameter, for instance the specific volume, the c/a ratio or the composition in an alloy, changes in such a way that two bands cease to overlap, producing a full valence band and an empty conduction band with an energy gap between them (see Fig. 4.1). A simple case is that due to the change in volume of a divalent metal. In *any* divalent metal, if the volume increases sufficiently, an s-like valence band will separate off from a p-like conduction band, the density of states going from the form of Fig. 1.13(b) to that of Fig. 1.13(c). The most favourable case is mercury,

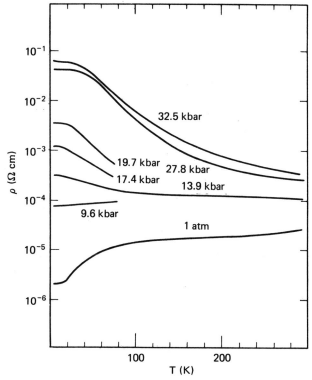

Fig. 1.14 Resistivity ρ of ytterbium as a function of temperature at various pressures (McWhan *et al.* 1969).

where we have already seen (Section 2.2.4) that the top of the valence band is s-like and expansion will increase the gap and not decrease it. There is no experimental method for appreciably increasing the specific volume of crystalline mercury, but experiments on liquid mercury at high temperatures show clearly that for densities below about $5 \, \mathrm{g \, cm^{-3}}$ the electrical properties are those of a semiconductor. These experiments are reviewed in Section 10.

The divalent metals Mg, Ca, Ba, Yb differ from Hg in the way illustrated in Figs. 1.3 and 1.4; the bottom of the upper band is s-like and compression will at first *decrease* the overlap, while in Hg it will increase it. Face-centred ytterbium under pressure has been investigated by McWhan *et al.* (1969) and Jullien and Jérome (1971). Figure 1.14 shows the resistivity of this element as a function of temperature at various pressures as determined by the former authors. Above 10 kbar, semiconducting behaviour is shown, the resistivity tending to a constant value at low temperatures presumably because quite a small concentration of impurities gives rise to a degenerate impurity band, due to a high background dielectric constant (cf. Chapter 6). Calcium† (McWhan *et al.* 1969) and strontium (Drickamer *et al.* 1966, Drickamer and Frank 1973) show a similar effect.

---

† The band structure of this material under pressure has been investigated by Altmann *et al.* (1971).

For the face-centred cubic divalent metals considered here, although the general considerations described suggest that the band gap *on the* (111) *plane* will increase with pressure, the proof that a gap can actually occur between the top corners of the first zone and the bottom of the second is more complicated. According to Vasvari *et al.* (1967) and Johansen and Mackintosh (1970), this behaviour is related to the presence of empty d-states not too far above the Fermi level, which hybridize with s-states near the corner of the first zone and depress their energies. According to McWhan *et al.* (1969), to obtain a true gap, spin–orbit coupling must be taken into account; without this there is a degeneracy keeping the bands in contact.

Confining ourselves to the model of non-interacting electrons, the transition is *second-order*, in the sense that there is no discontinuity in the value of n, the number of carriers. The band-gap should vary as $|a - a_0|$, where a_0 is the value of the lattice parameter a at the transition; the number n of carriers should vary as $|a_0 - a|^{3/2}$ and the energy as $|a_0 - a|^{5/2}$. The conductivity of a *perfect* crystal at zero temperature should change from zero to infinity, but if a finite mean free path is introduced there will be no discontinuity in the conductivity. This conclusion is changed, however, when electron–electron interaction is taken into account as in Chapter 4.

Many other metal–insulator transitions can be treated formally as band-structure transitions. For instance, for the semimetal bismuth, for which the Fermi surface consists of pockets of electrons and holes, studies of both the residual resistance and de Haas–Shubnikov oscillations show that the Fermi surface shrinks to zero at 25 kbar (Souers and Jura 1964, Balla and Brandt 1965, Itskevich and Fisher 1968). Iodine becomes metallic under a pressure of 160 kbar, without a change in the molecular structure and also with no change in volume (Drickamer *et al.* 1966). These authors report that the optical band gap and that deduced from the temperature dependence of the conductivity correspond well, and that for pressures giving metallic conduction $d\rho/dT$ is positive. Jayaraman *et al.* (1970b) observe that SmTe becomes metallic under pressure, the 4f-band apparently being pushed into the conduction band, again without discontinuous change in volume. On the other hand, EuO shows a discontinuous contraction by about 5% in volume at 300 kbar, thereafter showing a silvery lustre, indicating metallic behaviour (Jayaraman 1972); there is a further transition from rocksalt to CsCl structure at 400 kbar. In showing a discontinuous emptying or partial emptying of the 4f levels, it resembles metallic Ce and Yb. The existence of a discontinuity will depend on whether a tangent can be drawn to the free-energy–volume curve with two points of contact, as in Fig. 1.15. We should expect the free energy to increase less rapidly while the f-band is emptying, but whether this will lead to a two-phase region cannot be predicted a priori. Ramirez and Falicov (1971) (see also Kiwi and Ramirez 1972, Goncalves da Silva and Falicov 1972) first attempted a numerical description for cerium by introducing an interaction energy between the localized f-electron and the conduction electrons. A detailed discussion of more-recent developments is given by Singh and Ramesh (1985). The most striking feature is the transition at say 400 K, without change of

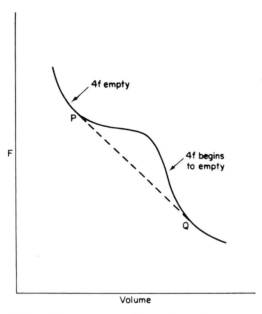

Fig. 1.15 Suggested plot of free energy against volume for a rare-earth metal. The electrons in the f-band are emptying into the conduction band for volumes as shown.

structure, of face-centred γ-cerium to α-cerium at a pressure of about 1 GPa with a volume change of order 20%. Of the many models proposed, we mention that of Coqblin and Blandin (1968) and Allen and Martin (1982), who assume that cerium is a "Kondo metal" (Chapter 3, Section 9) with a Kondo temperature very sensitive to volume, rising from 100 K in γ-Ce to 1000 K in α-Ce. It is not clear, however, why two minima in the free-energy curve should occur as in Fig. 1.15, though the calculations of Allen and Martin do show the required minimum.

Kirk *et al.* (1972) show that at 6.5 kbar there is a discontinuous change in the reflectivity of SmS, suggesting a metal–insulator transition, which was later observed by Bader *et al.* (1973) with an 8% volume change (see also Chapter 4 below and Jayaraman *et al.* 1970a, b, Maple and Wohlleben 1971).

The transitional-metal oxides are discussed in Chapter 6. It is suggested that the transition observed in Ti_2O_3 with increasing temperature is in this category.

5 The field round an impurity; screening and virtual bound states

If a positive charge ze is immersed in a degenerate electron gas, the Coulomb field is screened by the electrons. The screening was first estimated by the present author using the Thomas–Fermi approximation (Mott 1936, Mott and Jones 1936, p. 86), and by this method one finds for the potential energy of an electron

$$V(r) = -\frac{ze^2}{r}e^{-qr}, \tag{21}$$

where†

$$q^2 = \frac{4me^2}{\hbar^2} \left(\frac{3N_0}{\pi}\right)^{1/3}. \tag{22}$$

This formula for the screened Coulomb field has been used for many applications; it gives, for instance, a fairly satisfactory description of the residual resistance due to a small concentration of Zn, Ga, etc. in monovalent metals (Mott and Jones 1936, p. 293). Friedel (1956) points out, however, that the use of the Born approximation gives too large a value of the scattering, and better values are obtained if one calculates the phase shifts exactly.

More recent work has shown, however, that an exponential decay of the screening potential as in (21) is not correct, and that round any scattering centre the charge density falls off as $r^{-3} \cos 2k_F r$. This we shall now show, by introducing the phase shifts η_l defined as follows (cf. (13)). Consider the wave functions F_l of a free electron in the field of an impurity. These behave at large distances from the impurity as (Mott and Massey 1965)

$$F_l \sim Ar^{-1} \sin(kr - \tfrac{1}{2}l\pi + \eta_l) P_l(\cos\theta). \tag{23}$$

In order to obtain the charge density, we choose A such that an electron is in a sphere of radius R, so that $2\pi A^2 R = 1$, and take the sum

$$\sum_l (2l+1)(|F_l|^2 - |f_l|^2),$$

where f_l is the solution corresponding to (23) in the absence of the impurity. The charge fluctuations are thus given by

$$\frac{1}{2\pi R}\sum_k \sum_l \frac{2l+1}{r^2}[\sin^2(kr+\eta_l-\tfrac{1}{2}l\pi) - \sin^2(kr - \tfrac{1}{2}l\pi)].$$

We can evaluate the sum by multiplying by $R\,dk/\pi$ and integrating, which leads to the expression

$$\frac{1}{2\pi^2 r^2}\sum_l (2l+1) \int_0^{k_F} \sin(2kr + 2\eta_l - l\pi)\sin 2\eta_l \, dk.$$

A discussion of this formula for all values of l is given by Caroli (1967). We shall limit ourselves to the case where $l=0$ and where the phase shift η_0, due to a potential V_0 finite within a volume Ω_0, is small. In this case we write

$$V_0 = \int V \, d^3x / \Omega_0,$$

and then (Mott and Massey 1965, Chap. 5)

$$\eta_0 = 2mk\Omega_0 V_0 / 2\pi\hbar^2.$$

---

† Ziman (1964a, p. 130) gives the equivalent formula $q^2 = 4\pi e^2 N(E_F)$, which is the same as (22) for the free-electron case, $N(E)$ being defined for a single spin direction.

A simple integration gives for the change $\delta\rho$ in charge density (cf. Ziman 1964a, p. 137)

$$\delta\rho = \frac{\eta_0}{2\pi r^3}\left(\cos 2kr - \frac{1}{2kr}\sin 2kr\right). \tag{24}$$

Caroli's analysis shows that the fall-off as $\cos(2k_F r + \eta)/r^3$ is valid under more general conditions. This formula is used in Chapter 3 to derive the so-called RKKY interaction between magnetic moments embedded in a metal.

The phase shifts η_l obey the Friedel sum rule† (Friedel 1954)

$$z = \frac{2}{\pi}\sum_l (2l+1)\eta_l(k_F), \tag{25}$$

where the $\eta_l(k_F)$ are the phase shifts at the Fermi level and z is the charge on the scattering centre (z is, for example, 1 for Zn in Cu, 2 for Ga in Ag). We see that, for $z \geqslant 1$, *small* phase shifts for all l are not possible. Particularly important as giving large phase shifts is the concept of the virtual bound state or resonance introduced by Friedel (1956) for transitional-metal atoms in a matrix such as Cu or Al. Owing to the repulsive term $l(l+1)/r^2$ in the Schrödinger equation, the wave functions penetrate into the impurity only over a narrow range Δ of energies, typically of width $\lesssim 1$ eV. In such a case we introduce the *local* density of states at the impurity given by

$$N_0(E) \sim 1/\Delta.$$

We also introduce the overlap energy integral

$$I_{sd} = \int \phi_d^* V \phi_s \, d^3x$$

between the d wave function on the transitional-metal atom and those of the conduction electron on a neighbouring atom; the two are related by the formula (Anderson 1961)

$$N_0(E) = \frac{1}{\pi I_{sd}^2 \Omega N_s(E)}. \tag{26}$$

We use this formula in Chapter 3 in a discussion of the Kondo effect.

The large phase shifts η_2 give a large enhancement of the resistivity when transitional metals are dissolved in other metals. A survey for solid metals is given by Friedel (1956), and for solutions of Fe and Co in liquid germanium and tin by Dreirach et al. (1972). The resonance will also enhance the electronic specific heat and the Pauli paramagnetism, but these quantities cannot be treated quantitatively without including correlation as shown in Chapter 3.

---

† This rule appears to be valid also when electron–electron interaction is taken into account (see Langer and Ambegaokar 1961).

6   The mean free path

6.1   The Boltzmann formulation

If an electric field F acts on an electron in a state with wave function of the form (7) then k increases according to the equation

$$\frac{dk}{dt} = \frac{eF}{\hbar}. \tag{27}$$

The easiest way to prove this is by invoking the conservation of energy; the rate of increase of the energy W is

$$\frac{dW}{dt} = \frac{dW}{dk}\frac{dk}{dt} = \hbar v \frac{dk}{dt},$$

where v is the group velocity of a de Broglie wave. Since the rate at which energy is transferred to the electron by the field is eFv, (27) follows from equating these two quantities.

In order to calculate the conductivity of a solid, we have to introduce the concept of the scattering of electrons by defects such as impurities or phonons, or by the absence of crystalline order. The problem can be approached in two ways. The first is the Boltzmann formulation in which an electron is considered to be in a state with wavenumber k and to be scattered into another state k' by the field of the defect. This method becomes unsatisfactory if the scattering is strong, so that the mean free path l is not large compared with π/k. Under such conditions, we make use of the Kubo–Greenwood formulation (Kubo 1956, Greenwood 1958), in which one starts with eigenstates for the electron in the field of the imperfect lattice, the mean free path appearing as the distance in which phase memory is lost. Both will be described in this chapter.

For the Boltzmann formulation we introduce, for each point on the Fermi surface, the relaxation time τ, defined so that any disturbance δf of the equilibrium Fermi function f decays according to the law

$$\frac{d(\delta f)}{dt} = -\frac{\delta f}{\tau}.$$

Then in the presence of a field F along the x-axis, when a steady current is flowing, the Fermi surface will be shifted in the direction of the field by an amount

$$\delta k_x = eF\tau/\hbar, \tag{28}$$

as shown in Fig. 1.16. The contribution to the current of an element of area dS of the Fermi surface is thus

$$\frac{2e}{8\pi^3} dS\, \delta k_x\, v \cos^2 \theta,$$

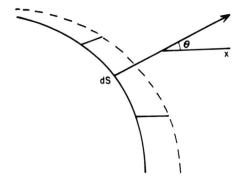

Fig. 1.16 Part of the Fermi surface in a metal displaced when a current is flowing.

where v is the group velocity of the electron normal to the surface and θ is as shown in Fig. 1.16. By (28), this becomes

$$\frac{1}{4\pi^3\hbar}\, dS\, e^2 F \tau v \cos^2 \theta.$$

τv is by definition equal to the mean free path l, so the conductivity is given by

$$\sigma = \frac{e^2}{4\pi^3\hbar} \int l \cos^2 \theta\, dS.$$

If l is constant and the Fermi surface is spherical with surface area S_F, this reduces to

$$\sigma = S_F e^2 l/12\pi^3\hbar. \tag{29}$$

Another way of obtaining the same result is the following. The Fermi distribution function f in the presence of a field F along the x-axis is given by

$$f_0(k) - \frac{df_0}{dE}\frac{dE}{dk}\frac{k_x}{k}\frac{eF\tau}{\hbar}, \tag{30}$$

where f_0 is the value of f in the absence of a field. The current j is given by

$$j = \frac{2e}{8\pi^3}\int f(k) v_x\, d^3k,$$

and, on substituting from (30) and writing

$$v_x = \frac{1}{\hbar}\frac{dE}{dk}\frac{k_x}{k},$$

we find

$$j = -F \int \frac{\partial f}{\partial E}\sigma(E)\, dE, \tag{31}$$

where

$$\sigma(E) = \frac{e^2}{4\pi^3\hbar} \int \left(\frac{k_x}{k}\right)^2 l\,dS \tag{32}$$

and, as before,

$$l = \tau v.$$

This form will be used in the next section for comparison with the Kubo–Greenwood formulation. $\sigma(E_F)$ is the conductivity at $T=0$ of an electron gas whose Fermi energy is E_F; (32) clearly reduces to the form (29) for a spherical Fermi surface over which the mean free path l is constant.

It remains to calculate l. We consider the case when the resistivity is due to impurities. Limiting ourselves to the case of a spherical Fermi surface, we suppose that there are N_0 scattering centres per unit volume, and that for each of these the differential cross-section for the scattering of an electron through an angle θ into the solid angle $d\omega$ is $I(\theta)\,d\omega$. Then

$$l^{-1} = N_0 \int I(\theta)(1 - \cos\theta)\,d\omega. \tag{33}$$

The factor $1 - \cos\theta$ takes account of the fact that, for the electrical resistance, small-angle scattering is less effective than large-angle. The proof is given in most textbooks on solid-state physics and will not be reproduced here.

If the scattering process is elastic, as it is for impurities or for disorder, then l given by (33) and the related relaxation time τ are the only quantities of importance. If, on the other hand, scattering is inelastic then two different times are relevant: that which determines the conductivity given by (33) with $\tau = l/v_F$, and that which does not contain the factor $1 - \cos\theta$, which is relevant to the heat conductivity. This leads to a breakdown of the Wiedemann–Franz law relating the two conductivities, particularly at low temperatures, when scattering by phonons is through small angles. For a discussion see Ziman (1964a, p. 386).

A discussion of the validity of formulae (29) and (33) has been given by Peierls (1955, p. 140). He shows that the condition is

$$\hbar/\tau < E_F,$$

or, setting $E_F = \frac{1}{2}mv^2$, $\tau = l/v$,

$$kl > 1/2\pi.$$

Ioffe and Regel (1960) were the first to point out that small values of l such that $kl < 1/2\pi$ are in fact impossible. This point is developed further in the next section. Moreover, when $kl \sim 1/2\pi$, which in metals is equivalent to the condition $l \sim a$ where a is the distance between atoms, the Boltzmann approximation becomes a bad one because k is no longer a good quantum number. This occurs particularly in non-crystalline materials and for impurity conduction.† In this situation we make use of the Kubo–Greenwood formula, which has universal validity; this will be described in the next section.

6.2   The Kubo–Greenwood formulation

We confine ourselves here to situations where the mean free path is due to elastic collisions with impurities or, in alloys, liquids and amorphous materials, with the non-periodic field in such materials. According to the Kubo–Greenwood formula, the current at temperature T is given by (31), where, instead of (32), $\sigma(E)$ is defined by the equation

$$\sigma(E) = \frac{2\pi e^2 \hbar^3}{m^2} |D_E|^2_{av} [N(E)]^2. \tag{34}$$

Here D_E is the matrix element

$$D_E = \int \psi_{l'}^* \frac{\partial}{\partial x} \psi_l \, d^3x; \tag{35}$$

ψ_l, $\psi_{l'}$ are wave functions for the energy E and the suffix "av" means an average over all states l, l' with this energy.

An elementary proof of this theorem is given by Mott and Davis (1979, Chapter 2). This depends on calculating $\sigma(\omega)$, the conductivity for frequency ω, and then making ω tend to zero. The main steps in the calculation are as follows. A perturbing field $Fe^{i\omega t}$ is applied to a degenerate electron gas. The perturbing energy is $eFxe^{i\omega t}$. The conductivity is given, for small ω, by the product of the following terms:

(i)   the number $N(E_F)\hbar\omega$ of electrons that can make the transition;

(ii)   the transition probability given according to Fermi's rule by

$$e^2 F^2 (2\pi/\hbar) |\langle l| \, x \, |l'\rangle|^2 N(E_F);$$

(iii)   the energy $\hbar\omega$ absorbed in each transition.

Here $\langle l| \, x \, |l'\rangle$ is the matrix element of x, and, writing

$$\langle l| \, x \, |l'\rangle = \frac{\hbar^2}{m\omega^2} \left\langle l \left| \frac{\partial}{\partial x} \right| l' \right\rangle,$$

we find

$$\sigma(\omega) = \frac{2\pi e^2 \hbar^3 \Omega}{m^2} \int |D_{av}|^2 \frac{N(E)N(E+\hbar\omega)}{\hbar\omega} \, dE,$$

---

† Under these conditions, we believe the factor $1 - \cos\theta$ to be unimportant, although this is queried by Morgan *et al.* (1985).

where Ω is the atomic volume and

$$D = \left\langle l \left| \frac{\partial}{\partial x} \right| l' \right\rangle.$$

On making ω tend to zero, formula (34) follows.

For large values of the mean free path and a spherical Fermi surface, (34) should lead to the same value (31) as the Boltzmann formulation. Edwards (1958) was the first to show that this was so, for the special case of weak scattering centres distributed at random in space; in this paper formula (33) for l was deduced. Mott and Davis (1979), taking l as the distance in which phase coherence is lost, give a proof that the Kubo–Greenwood formulation reduces to that of Boltzmann when $l > a$.

6.3   The case when $l = a$ (Ioffe–Regel hypothesis)

In Section 6.1 we referred to the Ioffe–Regel principle of a minimum value for the mean free path. If we think in terms of a tight-binding approximation then a mean free path shorter than the distance between atoms is obviously impossible; the strongest scattering that we can envisage is such that the phase of the wave function changes in a random way on going from one atom to another. Figure 1.17 shows a potential energy $V(x, y, z)$ introduced by Anderson (1958) that is convenient for discussing this kind of problem. It shows a crystalline array of potential wells, with a random potential V added to each, V lying between the limits $\pm \frac{1}{2} V_0$. The two diagrams illustrate the case when V_0 vanishes and when V_0 is large. The types of wave function envisaged are shown in Fig. 1.18, where (a) represents the case $l = a$, that is to say, where the phase changes in a random way from atom to atom. Instead of (8), we could write

$$\psi = \sum_n c_n e^{i\phi_n} \psi(r - a_n), \tag{36}$$

where c_n and ϕ_n are random numbers. If one sets $l = a$ in (29), one finds

$$\sigma = S_F e^2 a / 12\pi^3 \hbar, \tag{37}$$

which, if $S_F = 4\pi k^2$ and $k = \pi/a$, as for a half-filled band, reduces to

$$\sigma = e^2 / 3\hbar a. \tag{38}$$

For $a = 3\,\text{Å}$ this is equal to $2800\,\Omega^{-1}\,\text{cm}^{-1}$. This value of the conductivity we designate by σ_{IR}, IR standing for the Ioffe–Regel condition. If the number of electrons in the band, n_0 per atom, is much less than one then, since the conductivity is proportional to $n_0^{2/3}$ and the minimum mean free path to $n_0^{-1/3}$, σ_{IR} is reduced by $n_0^{1/3}$.

The Ioffe–Regel principle, in our view, is valid always for states in mid-band (in the sense that l cannot be less than a in the Anderson model) and true for elastic collisions in the sense that kl is not less than about π anywhere in a band.

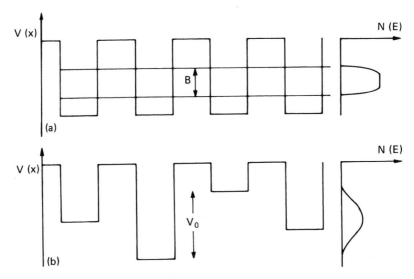

Fig. 1.17 Potential energy used by Anderson (1958): (a) without a random potential and (b) with such a potential. The density of states is also shown.

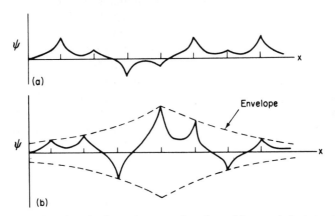

Fig. 1.18 Wave function ψ of an electron when $l \sim a$: (a) extended states; (b) weakly localized states.

But for inelastic collisions it is *not* true except in mid-band; several scattering processes can take place in one mean free path. An example is given in Chapter 10, Section 1.

We see that, if we use a wave function of the type (36) giving the conductivity (38), the question of the factor $1 - \cos\theta$ mentioned in the last section cannot arise.

An example of a very small "metallic" conductivity (Honig 1985) is quoted in Chapter 6 in our discussion of $(V_{1-x}Cr_x)_2O_3$ just on the metallic side of a metal–insulator transition as the temperature is increased. A value of σ, little dependent on temperature, of about $20\,\Omega^{-1}\,cm^{-1}$ is observed, less by a factor 100 than σ_{IR} for one electron per atom. It is believed on other grounds that here two bands overlap very weakly; it appears that the number of carriers must be of the order 10^{-6} per vanadium atom.

A beautiful example of the Ioffe–Regel principle is provided by the work of Milchberg *et al.* (1988), who determine the conductivity of aluminium at very high temperatures (electron temperature about 40 eV) without change of volume, by observing the reflection of an intense laser beam. The conductivity, $\sigma \approx 5 \times 10^{3}\,\Omega^{-1}\,cm^{-1}$, showing a plateau over a range of temperatures, corresponds well to σ_{IR} for three electrons per atom. That the electron gas is nearly non-degenerate will make little difference since $E_{F} \sim k_{B}T$. These authors explain the effect as "resistivity saturation", which in our view is identical with what is proposed here.

One might hope to obtain a better value of σ for $l=a$ by starting from the Kubo–Greenwood formula. Hindley (1970), Friedman (1971) and Mott (1972a) evaluate σ by summing (34) for each *pair* of atoms, using the tight-binding approximation. The number of pairs is $\frac{1}{2}zN$, where z is the coordination number and N the number of atoms. Each pair of atoms is treated as a bond, with its appropriate wave function. The matrix element δ for a single bond is taken as

$$\delta = maI/\hbar^{3}$$

(Holstein and Friedman 1968), so that

$$|D_{E}|^{2}_{av} = \tfrac{1}{2}zN\left(\frac{maI}{\hbar^{3}}\right)^{2},$$

assuming again that the phase varies in a random way, *but now from bond to bond.* I is as before the hopping integral (10) of the tight-binding method. Formula (34) then gives for the conductivity (for a simple cubic, for which a^{3} is the atomic volume)

$$\sigma = \frac{\pi e^{2}}{\hbar}za^{6}I^{2}[N(E_{F})]^{2}. \tag{39}$$

For a simple cubic lattice and a half-filled zone, we may set in the tight-binding approximation

$$N(E_{F}) \approx 1.75/2zIa^{3} \tag{40}$$

(Mott and Jones 1936, p. 85), so

$$\sigma = \frac{3\pi}{4z}\frac{e^{2}}{\hbar a}. \tag{41}$$

The assumption that $z=6$ seems reasonable for a random assembly of centres, and with $z=6$ this gives $\sigma = \pi e^{2}/8\hbar a$, a value very close to (38).

Equation (39) shows that the conductivity is proportional to $[N(E_F)]^2$, and $N(E_F)$ is proportional to the effective mass. On the other hand, formulae such as (41) do not contain the effective mass. This is because the integral D_E is inversely proportional to m_{eff}, as the analysis shows. However, if $N(E)$ is *less* than the free-electron value, as in a "pseudogap" (Section 16), no such cancellation occurs and the conductivity is proportional to $[N(E_F)]^2$. One may then write

$$\sigma = S_F e^2 a g^2 / 12\pi^3 \hbar, \tag{42}$$

where g is the factor introduced by Mott (1967) and defined by

$$g = N(E_F)/N(E_F)_{free}. \tag{43}$$

We assert also that this is so also in the Ioffe–Regel regime, when $l \sim a$, so that (42) remains valid here too.

The cancellation of g when $l > a$ in the equation for the conductivity was first discussed by Edwards (1962). He showed that if $l > a$, although (42) remains valid in principle (as follows from the Kubo–Greenwood equation (34)), one can write

$$l = l_0/g^2, \tag{44}$$

where l_0 is the value of the mean free path determined by first-order perturbation theory (with neglect of multiple scattering) and with no change in the density of states, as for instance in Ziman's (1961) theory of the resistivity of liquid metals (Chapter 10).

The arguments for Edwards' cancellation theorem are rather subtle, and we do not think that it is universally true (Mott 1989). For weak scattering, the relaxation time τ must be proportional to g^{-1}, by Fermi's golden rule. The mean free path l is given by the equation

$$l = v\tau,$$

where v is the group velocity of an electron at the Fermi surface. If the band is uniformly expanded then σ, given by $v = \hbar^{-1} dE/dk$, will also be proportional to g^{-1}, so (44) is valid. But for weak disorder we do not believe that uniform expansion necessarily occurs, and the density of states may even increase in mid-band (Van Oosten and Geertsma 1985). In Chapter 5, Section 4 we give another example. Our argument is that when $l \sim a$ perturbation theory cannot be applied, the mean free path has its minimum value and is no longer proportional to g^{-1}; also, since k is no longer a good quantum number, one can no longer write $v = \hbar^{-1} dE/dk$. So, in the neighbourhood of $l \sim a$, the factor g should be included in (42).

This factor g is of major importance in discussing the metal–insulator transition resulting from disorder. Using the model of Fig. 1.17, we take

$$g = \left(\frac{B^2}{B^2 + V_0^2}\right)^{1/2} \Big/ 1.75. \tag{45}$$

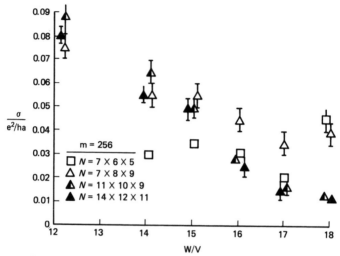

Fig. 1.19 Conductivity $\sigma/(e^2/\hbar a)$ as a function of W/V $(12V_0/B$ in our notation) calculated for a finite cube of varying size by Kramer *et al.* (1981). The cube side N is from 5 to 14 atoms.

This compares the width for a given value of V_0 with the bandwidth for the crystalline state. The factor 1.75 comes, as in (40), from comparing $N(E)$ in mid-band for a simple cubic, using the tight-binding approximation.

According to a calculation using the Born approximation (Mott and Davis 1979, p. 29, see (46)) we find that, with $z=6$, the Ioffe–Regel regime $l=a$ is reached when $V_0=0.6B$. We do not think that the Born approximation should be seriously in error; for repulsive fields it gives too much scattering, for attractive too little, and at mid-band the two should cancel. So at the Ioffe–Regel limit the factor $[1+(V_0/B)^2]^{1/2}$ should be about 1.1; if we include the factor 1.74 then g^{-1} is about 2 at the Ioffe–Regel limit.

We do not know how quickly the factor g comes into play as V_0 approaches $0.6B$. Figure 1.19 shows some results of Kramer *et al.* (1981) on the conductivity calculated for finite cubes. Extrapolation to $V_0/B=0.6$ suggests that $\sigma/(e^2/\hbar a)$ must be about 0.2, instead of the value 0.3 that is obtained for $g=1$.

In this book we suppose that, in the absence of quantum interference, we can use (42) and (45) when $\sigma < \sigma_{IR}$. This leads to self-consistent results in a number of experiments (see Chapters 5 and 10).

We discuss the value of g near the band edge in the next section.

Our factor g can arise either when a single band is broadened by disorder (Fig. 1.20a), when (45) is valid, or when two bands overlap, as in Fig. 1.20(b). In the latter case we *define* g through (42).

Equation (42) is valid only if multiple scattering is neglected; the effect of this, however, is of major importance at the transition, as we shall see in the following sections.

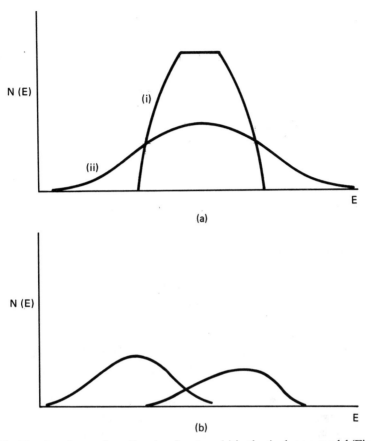

Fig. 1.20   Density of states in a disordered system: (a) in the Anderson model (Fig. 1.17) (i) shows the behaviour without disorder, simple cubic lattice, and (ii) that with disorder; (b) for two overlapping bands.

7   Disordered systems; localization, the Anderson transition and the mobility edge

Anderson (1958) was the first to show that in certain random fields "localization" of the one-electron wave functions can occur if the random component is large enough. To see what this means, we consider the three-dimensional potential-energy function $V(x, y, z)$ already illustrated in Fig. 1.17. This shows a crystalline array of potential wells, sufficiently far apart for the tight-binding approximation to be valid; if I is the overlap energy integral given by (10) then the bandwidth B is $2zI$. A random potential within limits $\pm\frac{1}{2}V_0$ is then introduced. If V_0 is small then (as we saw in the last section) the mean free path can be calculated from the Born approximation and is given by (Mott and Davis, 1979, p. 27)

$$\frac{l}{a} = 16\pi\left(\frac{I}{V_0}\right)^2 = \frac{4\pi}{z^2}\left(\frac{B}{V_0}\right)^2. \tag{46}$$

Thus if $z=6$ then $l=a$ when $V_0/B=0.7$. The wave function will then be as in Fig. 1.18(a), the phase varying in a random way from well to well as already described. According to Anderson, there exists a critical value $(V_0/B)_{crit}$, larger than that given above, for which diffusion is impossible at zero temperature; the eigenstates of the Schrödinger equation for all values of the energy then take the form shown in Fig. 1.18(b). The wave functions are then said to be "localized". They could be described by wave functions (cf. (36)) of the form

$$\psi = e^{-r/\xi} \sum_n c_n e^{i\phi_n} \psi_n(r-a_n), \tag{47}$$

where ξ is defined as the "localization length". If one has a Fermi gas of electrons and the wave functions for the Fermi energy E_F are localized then $\sigma(E_F)$ defined by (34) is zero, and thus no current can pass at absolute zero. When the temperature is slightly raised, conduction is by thermally activated hopping, which will be described in Section 15. Anderson (1970) has called a degenerate electron gas with localization in a random field of this kind a "Fermi glass".

The critical value $(V_0/B)_{crit}$ is difficult to calculate exactly; Anderson originally obtained for $z=6$ a value near 5, but subsequent calculations (Edwards and Thouless 1972) gave a value near 2. Numerical calculations by Schönhammer and Brenig (1973) confirm this. The value should be roughly proportional to the coordination number (Economou and Cohen 1972).

The most recent values of $(V_0/B)_{crit}$, together with the corresponding values of the factor g_c^{-1} are as follows:

	V_0/B	g_c^{-1}
Elyutin et al. (1984)	1.5	3.2
MacKinnon and Kramer (1981)	1.45	3.1
Pichard and Sarma (1981)	1.6	3.3
Schreiber (1987)	1.4	3.0

The effective coordination number should certainly be taken to have a value less than 6 for p- or d-functions, which are not spherically symmetric, giving smaller critical values, since substantial overlap between neighbours will occur only in certain directions. Experimental evidence that this is so is discussed later.

If the Anderson criterion is not satisfied then, as first pointed out by Mott (1966), since states are likely to become localized in the tail of a band, there exists a critical energy E_c (the "mobility edge")† separating localized from non-localized states (Fig. 1.21). The simplest definition of E_c in terms of the behaviour of the conductivity $\sigma(E)$ is as follows:

$$\sigma(E)=0 \quad (E<E_c),$$

$$\sigma(E)>0 \quad (E>E_c).$$

---

† The terminology is due to Cohen et al. (1969).

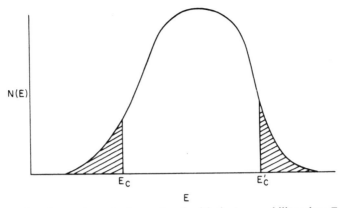

Fig. 1.21 Density of states in an Anderson band, with the two mobility edges E_c and E'_c.

If, by changing the composition of a material or due to the effects of strain or a magnetic field, the Fermi energy E_F can be made to cross the value E_c, a transition is expected from a metallic state, with finite values of $\sigma(E_F)$ (except actually when $E_F = E_c$) to a non-metallic state for which $\sigma(E_F)$ vanishes. The present author has called this an "Anderson transition". Calculations of the value of E_c for certain distribution functions for the energy states are due to Economou and Cohen (1972), Freed (1972), Abou-Chacra et al. (1973) and Abou-Chacra and Thouless (1974). Davies (1980) has calculated the position of the mobility edge in the conduction band of hydrogenated amorphous silicon, assuming that the localization is due to random dihedral angles, and finds that E_c lies about 0.2 eV above the bottom of the conduction band.

For the conductivity, again with neglect of multiple scattering, when E_F lies at E_c an equation similar to (42) should be valid, namely

$$\sigma = S_F e^2 a_E g_c^2 / 12\pi^3 \hbar, \tag{48}$$

where

$$\left(\frac{a_E}{a}\right)^3 = \frac{N(E_c)}{\langle N(E)\rangle}. \tag{49}$$

Mott (1985; see also Mott and Davis 1979, Chap. 2) maintains that $g_c = 0.3$ is valid also for $E = E_c$, wherever E_c may lie in the band, even near the band edge, as well as for transitions at mid-band. We need to define g for this case; we have to compare g, not for the same energy (in which case $N(E)$ for the crystal may be zero while for the amorphous state it will lie on the tail), but in another way, which will now be described. This argument is as follows (Mott 1985, 1988). Suppose the well depths have, say, a Gaussian distribution; then only those wells with depths below a given energy E contribute to the density of states at that energy, and from these a narrow band will result (Fig. 1.22). We argue that the treatment of Anderson can be applied to this band, and that for it localization (assuming $z = 6$)

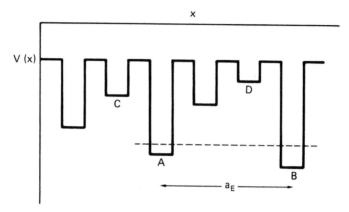

Fig. 1.22 Potential energy in the Anderson model, showing behaviour near a band edge. The dotted horizontal line shows the energy E considered. The wells marked A and B have levels below this energy E. The length a_E is the mean of distances between wells such as A and B.

will occur when $g \approx \frac{1}{3}$. For this reason we think that near the band edge $\sigma(E) \approx 0.03e^2/\hbar L_i$, where L_i is the inelastic diffusion length (see Section 10), the constant 0.03 being $\frac{1}{3}g_c^2$ as before.

Unless we are working in a tight-binding situation with a low coordination number, we believe that these values of V_0/B and g_c should be valid for all forms of disorder. What determines localization is the spread in the energies of atomic states—due to whatever cause; for nearest neighbours this must be about $3B$. It seems unlikely that localization can occur without a decrease in $N(E)$, an issue that is discussed further in Chapter 10.

For E_F near a band extremity, we find, using (49) for a_E, that σ_{min} drops by $(n/n_0)^{1/3}$ when there are n electrons and n_0 states in the band. We also note that localization occurs for less disorder for an array of p- or d-states because of the directional properties of the wave functions. Thus σ_{min} can be considerably greater for these functions. An example (VO_x) is given in Chapter 7.

With neglect of multiple scattering, Mott (1972a) supposed that if $E_F - E_c$ changed sign, causing a metal–insulator transition, or if a transition at mid-band resulted from a change of V_0, then the metallic conductivity would drop to a minimum value

$$\sigma_{min} = 0.03e^2/\hbar a_E, \tag{50a}$$

or

$$\sigma_{min} \approx 0.03e^2/\hbar a \tag{50b}$$

for E_F near mid-band. The factor 0.03 is the product of $\frac{1}{3}$, in the Ioffe–Regel expression (38), and g_c^2. σ_{min} has been referred to as the "minimum metallic conductivity". From the scaling theory of Abrahams et al. (1979) (see Section 13 below) and from many experiments (e.g. on doped semiconductors), it has become clear that no minimum metallic conductivity exists. In this book, following Bergmann (1983, 1984) and Mott and Kaveh (1985a, b), we describe this as caused by multiple scattering. This will be discussed in the next section.

First, however, we make some further remarks about the mobility edge E_c. We have stated that for energies below the mobility edge the wave function is described by a localization length ξ; ξ is believed to tend to infinity as $E_F \rightarrow E_c$, as $(E_c - E_F)^{-\nu}$; most of the evidence suggests that $\nu = 1$, although some earlier calculations (Lukes 1972, Freed 1972) suggested $\nu = \frac{2}{3}$. Mott (1983) has given arguments to show that ν cannot be less than $\frac{2}{3}$. However, detailed numerical work (Schreiber et al. 1988) leads to the value $\nu \approx 1.5$—a result that is difficult to reconcile with observations (apart from some results quoted in Chapter 5) of the *non-degenerate* electron gas in the conduction band of a semiconductor (Qui and Han 1982). This index is discussed further in Section 9 and Chapter 5).

The concept of a mobility edge has proved useful in the description of the non-degenerate gas of electrons in the conduction band of non-crystalline semiconductors. Here recent theoretical work (see Dersch and Thomas 1985, Dersch et al. 1987, Mott 1988, Overhof and Thomas 1989) has emphasized that, since even at zero temperature an electron can jump downwards with the emission of a phonon, the localized states always have a finite lifetime τ and so are broadened with width $\Delta E \sim \hbar / \tau$. This allows non-activated hopping from one such state to another; the states are delocalized by phonons. In this book we discuss only degenerate electron gases; here neither the Fermi energy at $T = 0$ nor the mobility edge is broadened by interaction with phonons or by electron–electron interaction; this will be shown in Chapter 2.

8 Weak localization; Bergmann's treatment

It was first pointed out by Bergmann (1983, 1984) that multiple scattering could produce a drop in the conductivity. The argument is illustrated in Fig. 1.23, which shows the Fermi surface of a metal in k-space, with an electric field along the k_x axis. An electron in a state represented by A, moving in the direction of the field, can be scattered through 180° to the point B, moving in the opposite direction, either directly or by multiple scattering events such as ACDB. In the latter case two possible paths, ACDB and AC'D'B are possible, and, if all scattering processes are elastic, these paths can interfere. Bergmann shows that this process reduces the conductivity by

$$\frac{\Delta\sigma}{\sigma} = -\frac{C}{(k_F l)^2}.$$

The constant C is uncertain, but we give evidence that it is unity. However, if some collisions are inelastic, this interference does not take place. We therefore introduce the inelastic diffusion length

$$L_i = (D\tau_i)^{1/2},$$

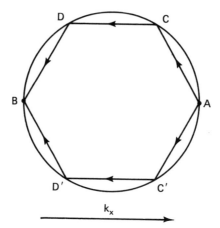

Fig. 1.23 Fermi surface in k-space, showing multiple scattering paths leading to quantum interference.

where D is the diffusion coefficient and τ_i the time between inelastic collisions. Then l/L_i is the chance that a collision is inelastic, so (as first suggested by Kaveh and Mott 1982) we set

$$\frac{\Delta\sigma}{\sigma} = -\frac{1}{(k_F l)^2}\left(1 - \frac{l}{L_i}\right). \tag{51}$$

The inelastic collisions can be either with another electron, when $\tau_i \propto T^{-2}$ (cf. Section 10) or with phonons, when above the Debye temperature $\tau_i \propto T^{-1}$.

These effects are described as "weak localization". Experimental evidence for the effects predicted for amorphous metals is described in Chapter 10. It will be seen that the effects are only significant when l is small, and thus typically for non-crystalline materials. It is remarkable that, as a consequence of (51), inelastic scattering *increases* the conductivity.

An equation of the type (51) has been obtained independently by other authors. The earliest (Kawabata 1981) is used in the next section. Fritsch *et al.* (1987), on the basis of the work of the Munich school, give

$$\Delta\sigma = \frac{e^2}{2\pi^2\hbar}(D\tau_0)^{-1/2}\left[\frac{2}{\pi} - \left(\frac{\tau_0}{\tau_i}\right)^{1/2}\right],$$

where τ_0 is the elastic scattering time, independent of T, and τ_i the inelastic scattering time. Apart from small variations in the constants, this is identical with (51).

Morgan *et al.* (1985) have developed an expression for the correcting term that is claimed to go beyond perturbation theory, and to be exact. However, it does not agree in the weak-perturbation limit with the formulae described above, giving a correction in $(kl)^{-2}F^{1/2}$, where F diverges logarithmically when $(kl)^{-1} \to 0$.

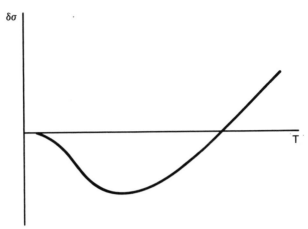

Fig. 1.24 Correction to the conductivity caused by spin–orbit scattering (Fukuyama and Hoshino 1981).

Economou *et al.* (1985) and Soukoulis *et al.* (1985, 1986, 1987) have used somewhat similar methods to calculate both the density of states, the mobility edge and the conductivity as a function of energy for the case of diagonal disorder; their work is limited to disorder parameters V_0 less than one-fifth of the bandwidth B, and is therefore relevant to the band tail.

A further effect is discussed in Chapter 5, namely that of spin–orbit scattering. This effect preserves the coherence of the two scattered waves but changes the constructive interference into a destructive one. The characteristic time τ_{so} is believed to be independent of T, and inversely proportional to Z^4, where Z is the atomic number. An analysis by Fukuyama and Hoshino (1981) shows that $\Delta\sigma$ depends on τ_{so}/τ_i, and leads to a behaviour as shown in Fig. 1.24. The negative part should be observed only for heavy metals.

9 Conductivities below σ_{min} near the Anderson metal–insulator transition

The effect leading to a term in L_i, described in Section 8, can be expressed by an equation first given by Kawabata (1981) and Kaveh and Mott (1982):

$$\sigma = \sigma_B\left[1 - \frac{C}{(k_B l)^2}\left(1 - \frac{l}{L_i}\right)\right],$$

where $\sigma_B = S_F e^2 l/12\pi^3\hbar$ is the Boltzmann conductivity. Mott and Kaveh (1985) extended this equation, for conductivities less than the Ioffe–Regel value for $l=a$, namely $\sigma_{IR} = \frac{1}{3}e^2/\hbar a$, by writing

$$\sigma = \sigma_{IR}g^2\left[1 - \frac{C}{(k_F a)^2 g^2}\left(1 - \frac{l}{L_i}\right)\right], \tag{52}$$

where g is defined by (43). The reason for adding g^{-2} to the second term is explained by these authors, who also remark that with this form σ obeys the scaling theory (Section 13). They also give reasons for expecting that it would be a fair approximation to extrapolate this equation to the metal–insulator transition.

For a half-filled s-band and if $l=a$, $k_F l=\pi$; we have seen that the transition takes place when $g=\frac{1}{3}$, so (52) gives a correct value for the vanishing of σ when $L_i=a$, if we set $C=1$. Our equation then becomes

$$\sigma=\frac{e^2}{3\hbar a}\left[1-\frac{1}{(\pi g)^2}\left(1-\frac{a}{L_i}\right)\right].\tag{53}$$

In the case $L_i=\infty$, that is $T=0$, it is clear that σ goes linearly to zero with decreasing g. For finite L_i we see that, when $(k_F a g)^{-2}=1$, $\sigma=\sigma_0$, where

$$\sigma_0=0.03e^2/\pi L_i.\tag{54}$$

For $L_i>\xi$, where ξ is the localization length, we may write

$$\sigma=0.03e^2/\hbar\xi,\tag{55}$$

with the same numerical factor, since from (52) it is clear that, when $L_i=\infty$, σ goes linearly to zero as g decreases, and when $L=\xi$ the numerical factor cannot change.

The factor 0.03 is deduced here for s-states and a half-filled band, and simple diagonal disorder (Fig. 1.17). We believe that it has much wider validity, and can be applied, at any rate in a theory of non-interacting electrons (as for instance in a semiconductor) to any form of disorder or for p- or d-states. Our confidence depends on the success of the scaling theory of Abrahams et al. (1979), which will be outlined in Section 13.

It should be noted that σ_0 given by (54) is also the pre-exponential factor in the conductivity when E_F lies below E_c, so that

$$\sigma=\sigma_0\exp\left(-\frac{E_c-E_F}{k_B T}\right),$$

as in non-crystalline semiconductors and in the ε_2 regime in impurity conduction (Chapter 6). This is on the assumption that the current path is at E_c. This has been queried (see Section 7 above and Overhof and Thomas 1989, Mott 1988).

10   The inelastic diffusion length

The concept of the inelastic diffusion length was introduced in Section 8, and is of major importance in discussions of quantum interference. It is defined by the equation

$$L_i=(D\tau_i)^{1/2},\tag{56}$$

where D is the diffusion coefficient and τ_i the time between inelastic collisions. In this section, then, we discuss τ_i. At low temperatures the Landau–Baber theory (Chapter 2, Section 6) would give

$$\frac{1}{\tau_i} \sim \left(\frac{k_B T}{E_F}\right)^2 \frac{\hbar}{ma^2},\tag{57}$$

so L_i^{-1} is proportional to T^ν, with $\nu = 1$. Moreover, as $D \to 0$ at the transition, L_i^{-1} becomes large.

In Chapter 2, Section 6 we point out that, according to the calculations of Kaveh and Wiser (1986), for an amorphous material the index ν may be between 1 and $\frac{3}{4}$.

At higher temperatures, scattering by phonons, always an inelastic process, will in most cases determine L_i. For temperatures above the Debye temperature Θ_D,

$$\frac{1}{\tau_i} \sim \frac{k_B T}{M\omega a^2}, \qquad \omega = \frac{k_B \Theta_D}{\hbar},\tag{58}$$

with a constant depending on the electron–phonon interaction, so

$$L_i \propto T^{1/2}.$$

Through the term l/L_i in (52) for the conductivity, an increase $\Delta\sigma$ in σ with $\Delta S \propto T^{1/2}$ is predicted; this has been observed for amorphous metals (Chapter 10). This arises because the probability of scattering by a phonon of energy $\hbar\omega$ should be proportional to ω, so for $T < \Theta_D$ we expect

$$\frac{1}{\tau} \propto \int \omega^3 \, d\omega \sim \omega_{max}^4,$$

and since $\omega_{max} \propto T$ this gives $\tau^{-1} \propto T^4$ and $L_i^{-1} \propto T^2$. In an intermediate region between this behaviour and the $T^{1/2}$ form, a value of L_i proportional to T is expected, and has been observed in metallic glasses (Chapter 10).

That all scattering by phonons is inelastic has been disproved by Afonin and Schmidt (1986); see Chapter 10, where we maintain none the less that $L_i = a$ in liquids, all collisions being inelastic, so the quantum-interference term is absent.

11   Cerium sulphide

Perhaps the earliest system to be described as an Anderson transition was cerium sulphide (Cutler and Mott (1969), on the basis of observations by Cutler and Leavy (1964)). The material in question can be written $Ce_{3-x}v_xS_4$, where v is a cerium vacancy, the vacancies being distributed at random. The field near a cerium vacancy repels electrons, because they are negatively charged. Variation of x, then, changes the number of electrons and the number of scatterers. Figure 1.25 shows some results on the conductivity. At that time the present author believed

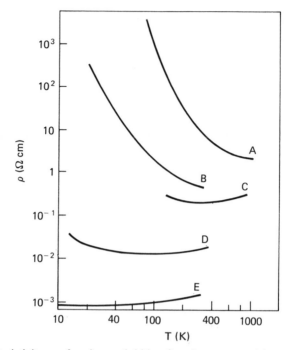

Fig. 1.25 Resistivity ρ of cerium sulphide of various compositions as a function of temperatures. The electron concentrations (in $10^{18}\,\mathrm{cm}^{-3}$) are as follows: A, 0.5; B, 5.2; C, 14; D, 83; E, 1420. From Cutler and Leavy (1964).

that a minimum metallic conductivity existed, and located this in the range 10^2–$10^3\,\Omega^{-1}\,\mathrm{cm}^{-1}$. In view of the considerations given here, it is likely that specimen D is metallic. Further work on this material at lower temperatures could be of interest.

12   Behaviour of the localization length

We now give evidence to show that, when E_F lies ΔE above E_c and ΔE is small we can define a length ξ, called the coherence length, which is the localization length for an energy ΔE below E_c. We show again that at $T=0$, i.e. when L_i is infinite,

$$\sigma \sim 0.03 e^2 / \hbar \xi, \tag{59}$$

$$\xi \sim a\left(\frac{E_c}{E_c - E_F}\right)^\nu, \quad \nu = 1. \tag{60}$$

All of these results neglect electron–electron interaction, which may profoundly affect them, though we do not believe that it affects (60).

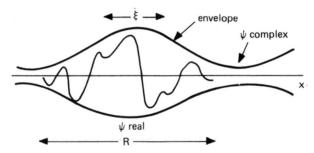

Fig. 1.26   Wave functions for energies just above E_c according to Mott (1983). ξ is the localization length for a function at the same energy below E_c.

The fact that σ goes continuously to zero at an Anderson transition, predicted first by the scaling theory of Abrahams *et al.* (1979), experiment (Chapter 5) and the argument given above, is difficult to reconcile with the Kubo–Greenwood formula (34), according to which σ is the integral of a squared quantity. According to Mott (1984) and Mott and Kaveh (1985b), this is to be understood by considering the form of the wave function ψ for an electron at an energy ΔE above E_c. According to these papers, over a range of length ξ it will differ little from the localized states at an energy ΔE below it. ξ is defined as the localization length of the latter (localized) function. The function for the extended state is therefore as in Fig. 1.26. In the peaks the wave function is real, and it is complex in the neighbourhood of the minima. The Kubo–Greenwood integral therefore has the form

$$\int \psi_1^* \frac{\partial}{\partial x} \psi_2 \, dx,$$

and over the range ξ the two functions ψ_1 and ψ_2 are real and identical, so the integral is of the form $[\psi^2]_{x_1}^{x_2}$, where x_1 and x_2 lie in the minima. It will be seen that σ must go to zero as E tends to E_c, and that the localization length and coherent lengths are the same for a given energy ΔE below and above E_c. A numerical investigation is given to justify the constant 0.03 in (59).

It is believed that a length such as ξ must tend to infinity at a second-order transition, just as the relevant lengths do at a critical or Néel point.

13   Scaling theory

The important paper of Abrahams *et al.* (1979) first made clear that the conductivity $\sigma(E)$ of a degenerate electron gas at zero temperature in a disordered environment must tend continuously to zero as the Fermi energy E_F tends to E_c. Abrahams *et al.* defined a dimensionless conductivity

$$G(L) = \frac{L\hbar}{e^2} \sigma(L), \tag{61}$$

where $\sigma(L)$ is the conductivity of a cube of side L. The proof depends on an argument of Thouless (1977), who showed that

$$G(L) = V(L)/W(L),$$

where $W(L)$ is the energy interval between quantized levels in the box, and $V(L)$ is the change in the energy levels resulting from a change in the boundary conditions—for instance from the vanishing of the wave functions ψ to the vanishing of $\partial\psi/\partial n$ differentiated normally to the surface. For $V(L)$ we may estimate as follows. A change in the boundary conditions will change the energy $\hbar^2 k^2/2m$ by

$$\frac{\hbar^2 k\, dk}{m} = \frac{\hbar^2 k}{mL}$$

if the mean free path l is infinite. However, for finite l only a layer of thickness l around the boundary is affected, so

$$V(L) \propto \frac{\hbar^2 k}{mL}\frac{l}{L}.$$

Thus, writing $N(E) \propto mk/\hbar^2$, we find

$$G(L) = \text{const} \times k^2 lL,$$

which is what we expect.

It is then argued that, if we fit together these blocks to form larger blocks, the only relevant quantity determining the new value of G is that for the smaller ones. Expressing this relationship as a differential equation, we find

$$\frac{d\ln G(L)}{d\ln L} = \beta(G(L)),$$

where $\beta(x)$ is a *universal* function of x. For large G, perturbation theory shows that

$$\beta = 1 - \frac{G_0}{G}, \tag{62}$$

and for small G

$$\beta = -e^{G_0/G},$$

where G_0 is a constant. Thus β will appear as in Fig. 1.27, changing sign at a universal value G_0 of G.

The important point is that in three dimensions β has a zero, so that near G_0, $d\ln G/d\ln L$ vanishes; therefore near this point G is constant (G_0) and thus

$$\sigma = G_0/\hbar L. \tag{63}$$

Scaling theory does not give the value of G_0, but in previous sections we found it to be approximately 0.03, and we deduce that this value must be universal.

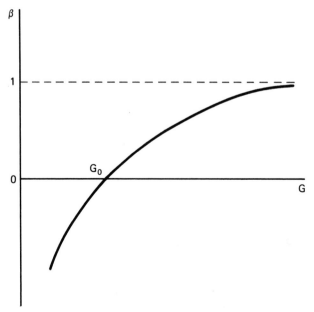

Fig. 1.27   β-function of scaling theory plotted against the dimensionless conductivity G.

Scaling theory has been criticised because of the large fluctuations in $\sigma(L)$ from specimen to specimen (when L is small). However, its success is showing that $\sigma \to 0$ for $L \to \infty$, and also that all states in two dimensions are localized, is widely recognized. But its conclusions when interaction is taken into account (Chapter 5) are not necessarily correct.

14   Proof of the Kawabata formula

In this section we give a proof of the Kawabata formula (52), following a method due to Kaveh (1984) and Mott and Kaveh (1985a, b). We assume that an electron undergoes a random walk, which determines an elastic mean free path l and diffusion coefficient D. If an electron starts at time $t=0$ at the point $\mathbf{r}_0$ then the probability per unit volume of finding it at a distance r, at time t, denoted by $n(\mathbf{r}, t)$ obeys a diffusion equation

$$\nabla^2 n(\mathbf{r}, t) + \frac{1}{D} \frac{\partial n(\mathbf{r}, t)}{\partial t} = 0, \tag{64}$$

the solution of which is

$$n(\mathbf{r}, t) = \frac{1}{(4\pi Dt)^{d/2}} \exp\left[-\frac{(\mathbf{r} - \mathbf{r}_c)^2}{4Dt} \right]. \tag{65}$$

Here d is the dimensionality of the system; $d=3$ in our case. For large values of t quantum mechanics must give the same result, so if $\psi(\mathbf{r}, t)$ is the wave function describing this system then

$$|\psi(\mathbf{r}, t)|^2 = n(\mathbf{r}, t). \tag{66}$$

We expand $\psi(\mathbf{r}, t)$ in terms of the eigenstates ϕ_n of the system, and obtain

$$\psi(\mathbf{r}, t) = N^{-1/2} \sum_n \phi_n^*(\mathbf{r}_0) \phi_n(\mathbf{r}) e^{i\omega_n t},$$

where N is the number of centres in a given volume and $\omega_n = E_n/\hbar$; this gives the required delta function $\delta(\mathbf{r} - \mathbf{r}_0)$ at $t=0$. From this, it follows that $|\psi|^2$, or $n(\mathbf{r}, t)$ given by (65), is

$$|\psi|^2 = N^{-1} \sum_{n, n'} \phi_n^*(\mathbf{r}_0) \phi_{n'}(\mathbf{r}_0) \phi_{n'}^*(\mathbf{r}) \phi_n(\mathbf{r}) e^{i\omega_{mn} t}.$$

Taking the time and space Fourier transforms of $|\psi|^2$, multiplying by $e^{-i\mathbf{q} \cdot \mathbf{r}_c}$ and integrating over $\mathbf{r}_0$, we obtain the following key relationship between the diffusion coefficient and the eigenstates:

$$|\langle \phi_n^* | e^{i\mathbf{q} \cdot \mathbf{r}} | \phi_{n'} \rangle|_{av}^2 = \frac{1}{\pi \hbar N_0(E_F)} \frac{Dq^2}{(Dq^2)^2 + \omega_{nn'}^2}. \tag{67}$$

Here the average is over all states separated by a fixed energy difference $\hbar\omega_{nn'}$, and $N_0(E_F)$ is the density of states per spin. This formula has been derived by various methods (MacMillan 1981, Kaveh and Mott 1981a, b, Imry et al. 1982, Lee 1982). For non-zero q, we find, making $\omega \to 0$, as in the derivation of the Kubo–Greenwood formula (34),

$$|\langle \phi_1^* | e^{i\mathbf{q} \cdot \mathbf{r}} | \phi_2 \rangle|_{av}^2 = [\pi \hbar N_0(E_F) Dq^2]^{-1}. \tag{68}$$

The functions ϕ are formed by the superposition of plane waves of given wavenumber $|\mathbf{k}|$ and with equal amplitudes but random phases for all directions of $\mathbf{k}$. For these plane waves, denoted in the absence of scattering by

$$|\mathbf{k}\rangle = \Omega^{-1} e^{i\mathbf{k} \cdot \mathbf{r}},$$

where Ω is the volume, we perturb them by the scattering and write

$$\psi_k = |\mathbf{k}\rangle + \sum_{q < 1/l} a(q) |\mathbf{k} + \mathbf{q}\rangle. \tag{69}$$

The upper limit on q of $1/l$ means that we do not include large-angle scattering but attempt to obtain the "best" wave function that retains a k-value. Our aim is to obtain the $a(q)$ and the conductivity, by feeding (69) into the Kubo–Greenwood formula, treating the large-angle scattering that determines ρ as in Mott and Davis (1979, p. 14).

We write an eigenstate as

$$\phi_n = \sum C_k^n \psi_k,$$

where the C_k^n are complex variables, and the sum is over all $\mathbf{k}$ for a given value of $|\mathbf{k}|$. We then have

$$\langle \phi_1 | q | \phi_2 \rangle = \sum \sum C_k^1 C_{k'}^2 \langle \psi_k | q | \psi_{k'} \rangle.$$

Here $\langle \phi_1 | q | \phi_2 \rangle$ etc. is written for $\langle \phi_1^* | e^{i\mathbf{q} \cdot \mathbf{r}} | \phi_2 \rangle$. If $|\mathbf{k} - \mathbf{k}'| \gg 1/l$ then the matrix element $\langle \psi_k | q | \psi_{k'} \rangle$ becomes negligible, so we restrict ourselves to $q < 1/l$. It is thus a reasonable approximation to write $k = k'$, and

$$\langle \phi_1 | q | \phi_2 \rangle = \sum_k C_k^{1*} C_k^2 \langle \psi_k | q | \psi_k \rangle.$$

Taking the square, we find

$$|\langle \phi_1 | q | \phi_2 \rangle|^2 = \sum \sum C_k^{1*} C_k^2 C_{k'}^1 C_{k'}^{2*} |\langle \psi_k | q | \psi_k \rangle|^2.$$

We argue that all the quantities $|C_k|$ must be identical, but that the C_k have random phases, so, on averaging, the summation reduces to

$$\sum |C_k^1|^2 \sum |C_k^2|^2,$$

which is unity by the normalization condition. Thus we find when $q \neq 0$,

$$|\langle \psi_k | q | \psi_k \rangle|^2 = [\pi \hbar D q^2 N_0(E_F)]^{-1}. \tag{70}$$

From this, we can deduce $|a(q)|$, noting that $\langle k | q | k \rangle = 0$, and we find

$$2a(q) = \langle \psi_k | q | \psi_k \rangle,$$

so that

$$|a(q)|^2 = [\pi \hbar D q^2 N(E_F)]^{-1} \tag{71}$$

To obtain the current, we make two assertions (cf. Kaveh 1985d). Neither the term in $\sum a(q) |\mathbf{k} + \mathbf{q}\rangle$ nor its product with $|\mathbf{k}\rangle$ make a contribution to the current in the Kubo–Greenwood formula. Thus the effect of this term on the conductivity arises from the normalization of the plane waves, written as

$$a_0 \left[|\mathbf{k}_F\rangle + \sum a(\mathbf{q}) |\mathbf{k}_F + \mathbf{q}\rangle \right]. \tag{72}$$

We see that

$$|a_0|^2 = \left[1 + \sum |a(q)|^2 \right]^{-1}$$

$$\approx 1 - \sum |a(q)|^2.$$

The sum can be replaced by an integral, and is

$$\left(\frac{L}{2\pi} \right)^d \int_{1/L}^{1/l} d^d q \, |a(q)|^2, \tag{73}$$

where d is again the dimensionality of the system and L is the size of the specimen, or the inelastic diffusion length, or the cyclotron radius $(c\hbar/eH)^2$, whichever is the smaller. The upper and lower limits are somewhat arbitrary; thus Berggren takes π/l, π/L, but we use $1/l$, $1/L$.

To evaluate $|a_0|^2$, we write, in three dimensions,

$$D = \frac{1}{3}\frac{\hbar k_F}{m}l, \quad N(E) = \frac{L^3 m k_F}{4\pi\hbar^2},$$

and find

$$\sum |a(q)|^2 = \frac{1}{\pi(k_F l)^2}\left(1 - \frac{l}{L}\right).$$

Thus for $|a_0|^2$ we obtain

$$|a_0|^2 = 1 - \frac{1}{2(k_F l)^2}\left(1 - \frac{l}{L}\right) \tag{74 a}$$

for the three-dimensional case, and

$$|a_0|^2 = 1 - \frac{1}{\pi k_F l}\ln\left(\frac{L}{l}\right) \tag{74 b}$$

for the two-dimensional case.

As regards the conductivity, in the Kubo–Greenwood formalism this will depend on $|a_0|^4$, so

$$\sigma = \sigma_B\left[1 - \frac{1}{(k_F l)^2}\left(1 - \frac{l}{L}\right)\right] \tag{75}$$

in three dimensions, and

$$\sigma = \sigma_B\left[1 - \frac{2}{\pi k_F l}\ln\left(\frac{L}{l}\right)\right] \tag{76}$$

in two dimensions. Here the constants 1 and 2 are rather uncertain, but both should be universal.

In the three-dimensional problem, it will be noticed from (71) that in $\sum|a_n|^2$ the density of states and the diffusion coefficient occur in the denominator, as they do also in the expression given by Kawabata (1981). If the disorder broadens the band, as will occur in the Anderson model if $V_0 > B$, then (75) should be modified to

$$\sigma = \sigma_B g^2\left[1 - \frac{C}{(k_F l)^2 g^2}\left(1 - \frac{l}{L}\right)\right], \tag{77}$$

the form given by (52). The constant C will be taken to be unity, which follows from the limits in (73): $1/L$ and $1/l$.

15 Hopping conduction

If states at the Fermi energy of a condensed electron gas are localized, two conduction mechanisms are possible.

(a) Excitation of the carrier to a mobility edge. If interaction with phonons is neglected then we expect the conductivity σ to be of the form

$$\sigma = \sigma_0 \exp\left(-\frac{E_c - E_F}{k_B T}\right),$$

with $\sigma_0 = 0.03 e^2 / \hbar L_i$. For modification of this equation on account of the displacement of the current path from the mobility edge resulting from delocalization by phonons see for instance Overhof and Thomas (1989) and Mott (1988).

(b) If wave functions are localized, so that $\langle \sigma(E) \rangle = 0$, then conduction at low temperatures is by thermally activated hopping. Every time an electron moves, it hops from one localized state to another, whose wave function overlaps that of the first state. Since the two states have quantized energies, the electron must exchange energy with a phonon (or possibly a spin wave) each time it moves. The hopping processes in which the electron obtains energy from a phonon are rate-determining. Hopping of this kind was first described by Miller and Abrahams (1960) in their theory of impurity conduction. They supposed that an electron on one occupied site would normally jump to a nearest site with energy ε_3 above it, where $\varepsilon_3 \sim [a^3 N(\varepsilon)]^{-1}$. The hopping probability is then of the form

$$\nu_{ph} \exp\left(-2\alpha R - \frac{\varepsilon_3}{k_B T}\right), \tag{78}$$

where ν_{ph} depends on the strength of the interaction with phonons and $\alpha = \xi^{-1}$, where ξ is the radius of the states. For large α the application of percolation theory to the problem of the most-favoured paths has been widely discussed (see e.g. Pollak 1972), but is outside the scope of this book. If α is small, as it is for $E = E_F$ in a system near an Anderson transition, then each localized state will overlap a large number of others. The factor $e^{-2\alpha R}$ may then be neglected over a volume of order ξ^3, and the electron will normally jump to the site in this volume with the smallest value of ε_3, which should be given by

$$\varepsilon_3 \sim \frac{(a\alpha)^3}{a^3 N(E)} = \frac{\alpha^3}{N(E)}. \tag{79}$$

Figure 5.2 shows some results of Davis and Compton for various forms of conduction in doped germanium, and the drop in ε_3 after an initial rise with decreasing a may perhaps be ascribed to the mechanism giving rise to (79), though other explanations (e.g. multiple hopping) have been proposed (cf. Pollak and Ortuno 1985).

At sufficiently low temperatures, under all circumstances where $N(E_F)$ is finite but states are localized near the Fermi energy, we expect the phenomenon of variable-range hopping to set in. If interaction between electrons is not taken into account then the conductivity will follow the formula

$$\sigma = A \exp\left[-\left(\frac{T_0}{T}\right)^{1/4}\right], \tag{80}$$

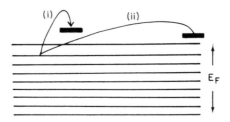

Fig. 1.28   The mechanism of hopping conduction. The electron jumps from a state below the Fermi level to a nearby state (i) or distant state (ii).

(Mott 1968, 1969, Ambegaokar *et al.* 1971, Pollak 1972). Here

$$k_B T_0 = 1.5/\alpha^3 N(E), \tag{81}$$

where $N(E)$ is the density of states at the Fermi level and α denotes the rate of fall-off of the envelope of the wave function in Fig. 1.18 ($\phi \sim e^{-\alpha r}$). The constant A, depends on the assumptions made about electron–phonon interaction. The behaviour from which (80) is deduced is illustrated in Fig. 1.28. An electron just below the Fermi level jumps to a state just above it, the energy required being written as ΔW. The farther it jumps, the greater is the choice of states that it has, and in general it will jump to a state for which ΔW is as small as possible. If it jumps a distance R, we can then write

$$\Delta W \sim [\tfrac{4}{3}\pi R^3 N(E)]^{-1},$$

so the conductivity is proportional to

$$\exp\left[-2\alpha R - \frac{3}{4\pi k_B T R^3 N(E)}\right]. \tag{82}$$

The formulae (80) and (81) are obtained by maximizing this quantity.

Since α in (81) is the reciprocal of the localization length ξ, measurements of T_0 near to the (Anderson) metal–insulator transition can determine how ξ varies with $n_c - n$. Thus Castner and co-workers (Shafarman and Castner 1986, Shafarman *et al.* 1986) have made measurements in Si : P and found, as expected, that for systems just below the Anderson transition ξ varies as $(n_c - n)^{-\nu}$, with $\nu \sim 1$.

However, it is now realized that it is only near the transition that (80) is valid—and even this is not certain. Efros and Shklovskii (1975) showed that the Coulomb repulsion between carriers led to a form of hopping for which

$$\sigma = \sigma_0 \exp\left[-\left(\frac{T'_0}{T}\right)^\nu\right], \quad \nu = \tfrac{1}{2}, \tag{83}$$

where (Shklovskii and Efros 1984)

$$k_B T'_0 = 2.8 e^2 / \kappa a. \tag{84}$$

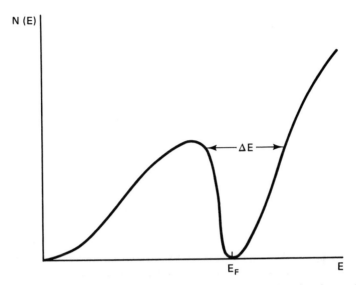

Fig. 1.29 Density of states $N(E)$ as a function of energy E in a disordered material in the localized regime, showing the Coulomb gap of width ΔE.

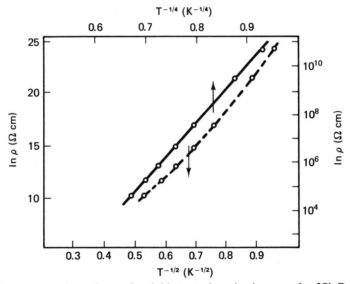

Fig. 1.30 Temperature dependence of variable-range hopping in a sample of Si : P: ———, plotted against $T^{-1/2}$; – – – –, plotted against $T^{-1/4}$ (Ionov *et al.* 1985).

The analysis first given by these authors depends on the proof that the *one-electron* density of states vanishes at E_F, showing the so-called "soft" Coulomb gap, illustrated in Fig. 1.29. This analysis is limited to systems far from the transition, where the overlap between wave functions is small. Here there is rather strong experimental evidence that $v=\frac{1}{2}$, particularly through the work of the Leningrad school. Figure 1.30 shows plots of the log resistivity of Si : P "without any special compensation" and $n=2.5 \times 10^{18}\,\mathrm{cm}^{-3}$ plotted against $T^{-1/4}$ and $T^{-1/2}$. The results are due to Ionov et al. (1985). It will be seen that $v=\frac{1}{2}$ is favoured, even though the concentration is quite near the transition, which is stated to be at $3.74 \times 10^{18}\,\mathrm{cm}^{-3}$. Zabrodskii (1980) found in Ge with $(5–7) \times 10^{17}$ donors cm^{-3} that the conductivity followed the hopping law with $v=\frac{1}{2}$. At the metal–insulator transition the concentration is about 10^{18}, so again the results are for specimens not far from the transition. Rentzsch et al. (1986) found $v=\frac{1}{2}$ in GaAs. Zabrodskii and Zinov'eva (1984) found $v=\frac{1}{2}$ in compensated n-Ge near the transition. On the other hand, there is some evidence that near the transition the index $v=\frac{1}{4}$ is valid. Thus Biskupski and Briggs (1987) and Biskupski et al. (1988) showed that when compensated barely insulating InP is subject to a magnetic field, driving it further from the transition by contracting the orbits, the hopping index changes from $\frac{1}{4}$ to $\frac{1}{2}$. Similar results for Si : P were obtained by Shafarman and Castner (1986) and Shafarman et al. (1986).

That the Coulomb gap might disappear, leading to $v=\frac{1}{4}$, can be understood in the following way. κ in (84) tends to infinity with ζ (see Chapter 5, Section 8). The width of the Coulomb gap is given by

$$\Delta E = [4\pi a \zeta^2 N(E_F)]^{-1},$$

which tends to zero as $\zeta \to \infty$. If

$$k_B T > \Delta E \sim \left(\frac{n-n_c}{n_c}\right)^2 \Big/ 4\pi a^3 N(E_F)$$

then we may expect that $v=\frac{1}{4}$ will be observed.

Apart from its effect on hopping conduction, there is direct evidence for a Coulomb gap in various materials. Thus Franz and Davies (1986) found evidence in sodium tungsten bronzes in the non-metallic state from the optical absorption coefficient. In a series of papers Whall et al. (1984, 1986, 1987) traced its influence in the conduction and thermopower of ferrites. Monroe et al. (1987) injected excess charge into p-type compensated GaAs, forming a layer of a metal–insulator semiconductor heterostructure. They found, in the delayed response to applied voltage, evidence for the Coulomb gap. Weng et al. (1983) found evidence for a Coulomb gap in their study of granular palladium films.

Whall et al. (1984, 1986) studied the electrical conductivity and thermopower of nickel ferrous ferrites ($\mathrm{Ni}_x\mathrm{Fe}_{3-x}\mathrm{O}_4$) with $0 < x < 0.9$. The thermopower for x between 0.1 and 0.8 tends to zero with temperature below 40 K, suggesting variable-range hopping (Whall et al. 1986). For higher temperatures they imply nearest-neighbour hopping. Whall et al. (1986, Fig. 3), in plots of $\mathrm{d}\ln(\sigma T)/\mathrm{d}(1/T)$ against T, showed an increase in the hopping energy round 300 K, which they ascribed to a Coulomb gap.

The Hall effect for hopping conduction is discussed in Chapter 5.

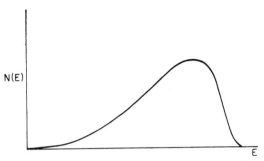

Fig. 1.31 Density of states $N(E)$ for a band with lateral disorder.

16 Pseudogaps and metal–insulator transitions

There have been many calculations of the density of states in various kinds of disordered material; we are concerned in this book mainly with narrow bands and thus with the tight-binding approximation. When the disorder is in the positions of the wells and not due to a random potential as in Fig. 1.17, the simple approximation (40) for $N(E)$ is not applicable. Some calculations have been made by Rousseau *et al.* (1970) and by Gaspard and Cyrot-Lackmann (1972, 1973). The type of density of states found by Gaspard and Cyrot-Lackmann is shown in Fig. 1.31, with low-energy tails due to groups of wells closer together than the average and an absence of Van Hove singularities. The minimum in the density of states predicted by Lifshitz (1964) is not found. These calculations do not distinguish between localized and non-localized states. Some rough estimates of the range of localized states for simple models have also been made (Mott 1970, Mott and Davis 1979). *Two* kinds of disorder can lead to localization. One is the random positions of the atoms, and this we believe gives a very narrow range of localization in liquids or glasses when the atomic functions are s-like, as in liquid rare gases (Mott and Davis 1979, Chap. 5). The other is the random orientation of p- or d-like functions, which should give a wider range of localization, perhaps a few tenths of an electron volt in an amorphous semiconductor.

For the subject matter of this book, it is of particular interest to consider the situation for a non-crystalline system analogous to that of crystalline ytterbium or strontium under pressure, namely that when a valence and conduction band are separate or overlap slightly. If the degree of overlap can be changed by varying the mean distance between atoms, the composition or the coordination number then a metal–insulator transition can occur. Many examples will be discussed in this book, particularly amorphous films of composition $(Mg_{1-x})_2(Bi_x)_3$, liquid mercury at low densities, and liquid tellurium alloys in which the coordination number changes with temperature. The transition is, we believe, of Anderson type.

If a conduction and valence band overlap slightly then a "pseudogap" or minimum in the density of states (Fig. 1.32) is expected, as first suggested by Mott (1966). As long as the overlap is small, one would expect the density of states, all

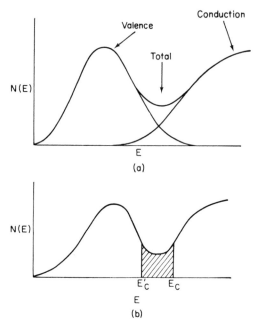

Fig. 1.32   (a) Overlapping bands forming a pseudogap, with localized states at the Fermi energy. (b) Total density of states, with localized states shaded.

localized, to be the sum of the contributions by the two bands. As the overlap increases, a point will be reached at which states at E_F are no longer localized. At this point a metal–insulator transition of Anderson type is expected. For weak localization, conduction by hopping according to (80) and (83) will occur. For extended states just above the transition, the conductivity will behave in a manner that we shall now discuss.

As in previous sections, we introduce the Mott g-factor, though here we must define it as being proportional to the density of states at the Fermi level, and normalized so that

$$\sigma = S_F e^2 a g^2 / 12 \pi^3 \hbar$$

without quantum interference as in (37). Putting $S_F = 4\pi k^2$, $k = \pi/a$, this gives

$$\sigma = g^2 e^2 / 3\hbar a.$$

We emphasize that the use of g in these equations may be justified only if $l \sim a$, because of the Edwards cancellation theorem (Section 6). We should expect a metal–insulator transition to occur for some value of g in the neighbourhood of $\frac{1}{3}$. For several liquid systems there is experimental evidence that the interference term in (52) is absent. Thus for liquid TeTl alloys, with variation of composition and temperature, for σ less than the Ioffe–Regel value $e^2/3\hbar a$, the conductivity is proportional to the square of the Pauli paramagnetic susceptibility and then to g^2. These results are due to Cutler (1977). Warren (1970a, b, 1972a, b) examined the Knight shift in liquid In_2Te_3 and Ga_2Te_3, from which he deduced g. He found σ to be proportional to g^2, g decreasing with decreasing temperature until g reaches a value of about 0.2; it then drops more rapidly, suggesting that localization occurs. These results are discussed further in Chapter 10.

Mott (1985, 1989a) proposed that in liquids all collisions are inelastic, so that $l = L_i$ and the interference term in (52) vanishes. Thus, in a sense, σ_{min} exists for liquids (see Chapter 10). We have to ask, however, whether localization exists under these conditions. The same problem exists in certain amorphous solids, where strong phonon interaction leads over a certain temperature range to the relation $L_i \sim a$. We find that, as the energy drops into a region of small density of states, τ_i increases and the normal localization condition again becomes valid.

Pseudogaps in liquids are discussed in Chapter 10. Examples of pseudogaps in solids ($Mg_{2-x}Bi_{2+x}$) are also described in Chapter 7.

2

Interacting Electrons

1 Introduction

The electrons in a solid interact both with one another and with the lattice vibrations. A theme of this book is the effect of the interaction between electrons in inducing magnetic moments and metal–insulator transitions. Interaction with phonons also has an important effect, particularly in some transitional-metal oxides. In this chapter both kinds of interaction are introduced.

2 Interaction with phonons and polaron formation

The phonons (lattice vibrations) in a solid or liquid conductor have the following effects on mobile electrons or holes.

(a) In metals or semiconductors they can lead to transitions from a state with one value of the wavenumber k to another, thus being responsible for an important part of the electrical resistivity. The collisions are inelastic, a phonon being absorbed or generated. This is described in most textbooks and will not be treated here.

(b) They can give energy to electrons in donor centres, thus exciting them into the conduction band, and conversely receive energy when electrons are trapped.

(c) In amorphous semiconductors and "Fermi glasses" they can exchange energy with electrons, moving them from one localized state to another (thermally activated hopping; see Chapter 1, Section 15).

(d) In amorphous semiconductors this process, hopping downwards, can shorten the lifetime of an electron in a localized conduction band state, and thus lead to delocalization (Dersch and Thomas 1985, Dersch *et al.* 1987, Mott 1988, Overhof and Thomas 1989).

(e) In general, interaction with phonons will lead to a moderate (by a factor of less than 2) enhancement of the effective mass of electrons within an energy $\hbar\omega$ of the Fermi energy of a metal (see below).

(f) Particularly for certain flat forms of a Fermi surface, they are responsible for a "Peierls distortion", introducing a gap over some or all of the surface (Section 4).

(g) They can interact with electrons or holes in a narrowband semiconductor or semimetal, forming polarons or bipolarons.

The behaviour of polarons is of importance for the oxides and similar materials to be discussed in this book, and we summarize some of their properties here, as applied chiefly to transitional-metal oxides. (For more detailed discussions see Mott and Davis (1979, Chap. 3), Austin and Mott (1969a, b), Appel (1968), Emin (1973, 1975) and Mott (1973b).

In ionic crystals a polaron is the region round an electron in the conduction band in which the material is polarized by the electron, as first proposed by Landau (1933). Round the electron, the field with which it interacts has a potential-energy function

$$V(r) = \begin{cases} -e^2/\kappa_p r & (r > r_p), \\ -e^2/\kappa_p r_p & (r < r_p), \end{cases} \tag{1}$$

where

$$\frac{1}{\kappa_p} = \frac{1}{\kappa_\infty} - \frac{1}{\kappa},$$

and κ_∞ and κ are the high-frequency and static dielectric constants. In principle, r_p is determined by minimizing the polaron's energy, which is made up of the potential and kinetic energies of the electron in the polaron,

$$-\frac{e^2}{\kappa_p r_p} + \frac{\hbar^2\pi^2}{2m^* r_p^2},$$

and the energy required to polarize the medium, $\frac{1}{2}e^2/\kappa_p r_p$. m^* is the effective mass. If r_p is larger than the lattice parameter, the polaron is called a "large" or Fröhlich polaron and its properties depend on the dimensionless coupling constant α, defined by

$$\alpha^2 = \frac{e^2}{\kappa_p} \left(\frac{m^*}{2\hbar^3\omega_0} \right)^{1/2};$$

the treatment given here is then not applicable. (For a review of the properties of the Fröhlich polaron see Devreese (1972).) Approximations based on the use of the coupling constant (which does *not* contain the lattice parameter) break down when r_p approaches a. The polaron is called a "small" polaron when r_p has its limiting size (Bogomulov *et al.* 1968)

$$r_p = \frac{1}{2}(\pi/6N)^{1/3}, \tag{2}$$

where N is the number of sites per unit volume. Only the small polaron is of importance for the phenomena described in this book. Its energy is $-W_p$, where

$$W_p = \tfrac{1}{2} e^2 / \kappa_p r_p.$$

The small polaron has the following properties.

(a) At high temperatures $(T > \tfrac{1}{2} \Theta_D)$ it moves by "hopping", the frequency with which it "hops" from one site to another being

$$\omega \exp(-W_H / k_B T), \tag{3}$$

where

$$W_H = \frac{1}{4} \frac{e^2}{\kappa_p} \left(\frac{1}{r_p} - \frac{1}{R} \right) \tag{4}$$

and R is the distance from one site to another. W_H is the energy of the intermediate state, where thermal fluctuations have decreased the depth of the potential well (1) and produced an empty potential well on a neighbouring site so that an electron can resonate between the two wells. Equation (3) is valid only in the so-called "adiabatic approximation", which means that the electron can go backwards and forwards several times during the period when the two wells have the same depth. If this is not so then the prefactor ω should be replaced by

$$\pi^{1/2} I^2 / \hbar (W_H k_B T)^{1/2} \tag{5}$$

Here I is the transfer integral, of the form $I_0 e^{-\alpha R}$. Thus it is only for comparatively large values of R that a tunnelling factor $e^{-\alpha R}$ is expected in the mobility. Murawski et al. (1979) have examined the conductivity of oxide glasses containing transitional-metal atoms of mixed valence, in which the carriers form polarons; they come to the conclusion that the term $e^{-\alpha R}$ is usually appropriate, V_2O_3–P_2O_5 and WO_3–P_2O_5 being possible exceptions, the carrier moving by the adiabatic process.

(b) At low temperature $(T < \tfrac{1}{2} \Theta_D)$ the electron moves from site to site without the aid of thermal activation, the zero-point energy $\tfrac{1}{2} \hbar \omega$ taking the place of thermal vibrations. The hopping frequency is now (Holstein 1959)

$$\omega \exp(-W_H / \tfrac{1}{2} \hbar \omega). \tag{6}$$

This can be seen from elementary considerations. The distortion of the surrounding medium that leads to the intermediate state with energy W_H can be described by simple harmonic motion, with a parameter x for the displacement, a potential energy $\tfrac{1}{2} p x^2$ and a wave function of the form $\Psi = \text{const} \times \exp(-\alpha x^2)$. The probability $|\Psi|^2$ of a configuration with potential energy W_H is thus

$$\exp(-2\alpha W_H / \tfrac{1}{2} p),$$

and, putting in the constants, it is easily seen that $4\alpha / p = 1 / \tfrac{1}{2} \hbar \omega$.

A very similar situation is presented by the dynamic Jahn–Teller effect, in which a trapped electron that has distorted its surroundings hops between two configurations with a frequency given by (6); for a proof see Sturge (1967).

Formulae similar to (6) have wide application to the movement of a system through an activated intermediate state, and will be applied to a simplified description of the Kondo effect in Chapter 3. For polarons the effect is to allow band motion at low temperatures with well-defined wavenumber k but enhanced effective mass m_p given by

$$\frac{m_p}{m^*} = \frac{\hbar}{2m\omega R^2} \exp\left(\frac{W_H}{\frac{1}{2}k\Theta_D}\right). \tag{7}$$

The prefactor is about 3; m^* is the effective mass in the undistorted lattice. The small polaron, for $T < \frac{1}{2}\Theta_D$, will be scattered by phonons; its mobility when plotted against T^{-1} will be as in Fig. 2.1.

These formulae can lead to a large mass enhancement of order 10, even if W_H is of order $\frac{1}{2}\hbar\omega$, so that hopping is unlikely to be observed even at high temperatures (see an analysis of the problem by Sumi 1972). We shall call particles with considerable mass enhancement (about 10) but no observable hopping motion "heavy polarons". These have been studied particularly by Eagles (1969a, b, 1984, 1985) and Eagles and Lalouse (1984).

We shall in this book use the concept of a "degenerate gas" of small—or at any rate heavy—polarons. Clearly we should not expect these to be formed unless the number of carriers is considerably less than the number of sites. We also remark, as mentioned earlier, that in all metals, at temperatures less than Θ_D, phonons lead to a certain mass enhancement, of order less than 2. A treatment is given by Ashcroft and Mermin (1976, p. 520). This affects the thermopower; some results for an amorphous alloy ($Ca_{1-x}Al_x$) from Naugle (1984) are shown in Fig. 2.2. A theoretical treatment of the range between this situation and the polaron gas has not yet been given.

Up to this point we have discussed the formation of polarons in ionic crystals. Polarons of another type can also form in elements and other systems, such as the valence bands of alkali and silver halides, where the polarizability is not the relevant factor. In fact Holstein's (1959) original discussion of the small polaron was of this form. This kind of polaron is sometimes called a molecular polaron, and is illustrated in Fig. 2.3(a), and in Fig. 2.3(b) in the activated configuration of the atoms when the electron can move from one site to another. There is nothing analogous to the large polaron in this case; in three-dimensional systems either a small polaron is formed or there is little effect on the effective mass from interaction with phonons.

In Fig. 2.3 the dots represent the atoms in the Holstein model, and (a) shows the situation where an electron sets up a bond between two atoms, pulling them together. A physical example is a hole in the valence band of a solid rare gas (e.g. Xe), forming a "molecule" Xe_2^+.

To obtain the activation energy for motion, we plot in Fig. 2.4 the energy of the system against the displacement q; Aq^2 is the elastic energy. The displacement

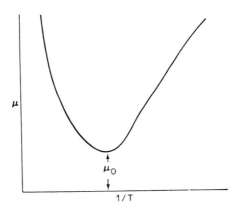

Fig. 2.1 Mobility μ of a polaron as a function of $1/T$. The minimum value, occurring when $T \sim \frac{1}{2}\Theta_\mathrm{D}$, is of order $\mu_0 = (\omega e a^2 / k_\mathrm{B} T) \exp(-W_\mathrm{H}/k_\mathrm{B} T)$.

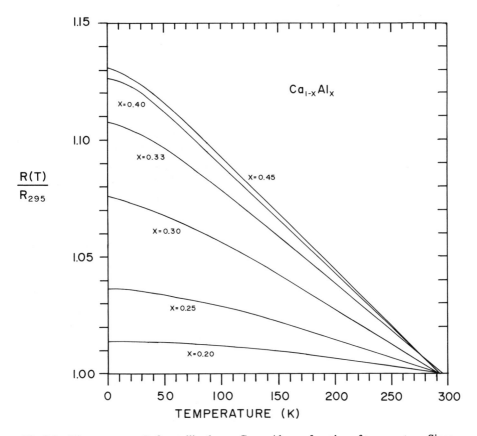

Fig. 2.2 Thermopower S of metallic glasses $Ca_{1-x}Al_x$ as a function of temperature. Since S is proportional to $k_\mathrm{B} T / E_\mathrm{F}$, these show an increase of $E_\mathrm{F}(=\frac{1}{2}\hbar^2 k^2 / m_\mathrm{eff})$ above about 200 K, suggesting a decrease in m_eff.

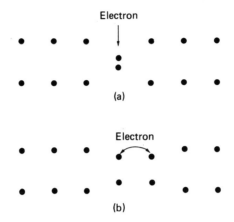

Fig. 2.3   (a) A molecular polaron and (b) the excited state that must be formed before an electron can hop from one site to another.

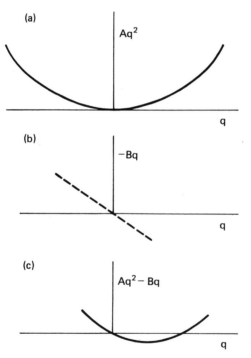

Fig. 2.4   Terms in the energy of an electron as a function of a configurational parameter q.

produces for the hole a "potential well" of radius about a, the distance between atoms, and depth $H = Bq$, where B is a constant. It is a well-known result of quantum mechanics that a bound state in such a well exists only if

$$m_{\text{eff}} H a^2 / \hbar^2 > 1, \tag{8}$$

so the well will not trap a carrier or lower its energy until q exceeds a critical value q_0; the energy of the system without the term Bq_0 will thus appear as in curve (b) of Fig. 2.4. We write the distortion energy as $-B(q - q_0)$ (though the energy behaves like $(q - q_0)^2$ near q_0). Then the total energy E is given by

$$E = Aq^2 - B(q - q_0), \tag{9}$$

which has a minimum value E_0 when $q = B/2A$, given by

$$E_0 = Bq_0 - \frac{B^2}{4A}. \tag{10}$$

The energy E will necessarily have this minimum, but its value at this point can be positive or negative; only in the latter case will a stable self-trapped particle (i.e. a small polaron) form. This is most likely to occur for large effective mass, and thus for holes in a narrow valence band or for carriers in d-bands. If the polaron is unstable then there is practically no change in the effective mass of an electron or hole in equilibrium in the conduction or valence band.

It will be seen that a barrier exists resisting self-trapping; this has been observed as a time delay by Laredo et al. (1981, 1983) for holes in AgCl, indicating a barrier height of 1.8 meV.

The carrier can move, either by excitation out of this self-trapped state into the conduction band, or by "hopping" to a neighbouring site. For hopping to occur, the interatomic distances between two adjacent pairs must be equal, as illustrated in Fig. 2.3(b). Then the electron can move freely from one such pair to the other. Possibly the electron can move backwards and forwards several times before the system relaxes; the transition is then adiabatic. Alternatively the chance of transfer in the time during which the configuration persists may be small.

At high temperatures the configuration of Fig. 2.3(b) is produced by thermal excitation, and the motion is by "thermally activated hopping", with mobility proportional to a factor of the type (3). To find W_H, we must ask what displacement of the kind shown in Fig. 2.4(b) has the smallest energy. This will be so when both displacements are equal to $B/4A$, as may easily be verified. In (10) E_0 then takes the form

$$W_p = \tfrac{1}{2} B^2 / A,$$

W_p being the polaron energy; we find the hopping energy

$$W_H = \tfrac{1}{2} W_p.$$

The hopping energy is thus one-half of the energy released when a polaron is formed. For this so-called adiabatic case, when the electron goes backwards and forwards several times during the period of excitation, the chance per unit time that the electron will have moved from one site to another after the system has relaxed is given by (3), where ω is the attempt-to-escape frequency. The diffusion coefficient D is thus

$$D = \tfrac{1}{6}\omega a^2 \exp\left(-\frac{W_H}{k_B T}\right), \tag{11}$$

and the mobility μ, from the Einstein relation $\mu = eD/k_B T$, is

$$\mu = \frac{ea^2\omega}{6k_B T} \exp\left(-\frac{W_H}{k_B T}\right). \tag{12}$$

For the non-adiabatic case we must, as before, replace ω by (5).

In crystalline materials a characteristic of polaron motion is a difference between E_σ, the activation for conduction, and E_S, that for the thermopower written as $S = (k_B/e)(E_S/k_B T + \text{const})$. We expect that $E_\sigma = E_c - E_F + W_H$ and $E_S = E_c - E_F$, where E_c is the extremity of the band in which the carriers move.

Finally we mention the important predictions of Friedman and Holstein (1963) for the Hall effect when charge transport is by polaron hopping. At first sight it is not obvious that, for hopping transport, there should be any Hall effect; but the authors quoted show that if we take three sites as in Fig. 2.5 so that a carrier can go from A to C either directly or via B, and if thermal activation is such that all three sites have the same energy, then there is an interference between the two paths, leading to an activated Hall mobility. An important conclusion is that the sign of the Hall effect is negative (n-type), whether the carriers are electrons or holes, so long as they are in s-like orbitals. Emin (1977a, b) discussed what happens if they are p-like; under certain conditions, electrons are then predicted to give positive, and holes negative, values of the Hall coefficient.

The analysis of Friedman and Holstein shows that, in the non-adiabatic case, the Hall mobility μ_H varies with temperature as

$$\mu_H = T^{-3/2} \exp\left(-\frac{W_H}{3k_B T}\right).$$

For the adiabatic case the activation energy can be smaller.

The polaron radius, if greater than (2), will be very sensitive to m^*—or, more exactly, to the bandwidth of the undistorted lattice. A particularly striking effect is that in materials like NiO doped with lithium, where the carriers are Ni^{3+} ions and in which the "hole" moves from one Ni^{2+} ion to another. The mass enhancement for a free carrier is rather small (about 5), while a bound carrier hopping round the Li^+ ion on the sites available to it behaves like a small polaron with an activation energy for motion (see Bosman and van Daal (1970) and Chapter 6 below).

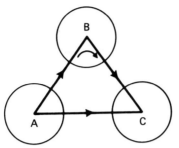

Fig. 2.5 The two paths between two sites A and C for polaron motion, which can give rise to a Hall effect.

A small polaron is normally bound to a donor (or acceptor) on a lattice site at a distance R with the binding energy

$$W = e^2/\kappa R, \tag{13}$$

where κ is the *static* dielectric constant. The usual formula for the binding energy of an electron in semiconductors, if m_p is the effective polaron mass, namely

$$W = m_p e^4/2\hbar^2 \kappa^2,$$

will be valid only if the Bohr radius

$$\kappa \hbar^2/m_p e^2$$

is greater than R. Since m_p/m is by hypothesis large, this will happen only when κ is very large. The clearest case when this occurs is in doped titanates, where $\kappa \sim 1000$ and a degenerate electron gas is observed for a density of carriers greater than $2 \times 10^{17}\,\mathrm{cm}^{-3}$. These are discussed in Chapter 4.

When a polaron can move between two sites of equal energy, the state of the system (electron + phonons) will split into two states of even and odd parity, separated by an energy $\Delta E \sim \hbar^2 \pi^2/m_p a^2$, which, when m_p is large, may be quite small. The polarizability of this system will lead to the formation of a moment in a field F of order

$$\left.\begin{array}{ll} e^2 F a^2/\Delta E & (k_B T < \Delta E), \\ e^2 F a^2/k_B T & (k_B T > \Delta E). \end{array}\right\} \tag{14}$$

For a discussion of the energy loss in such a situation see Austin and Mott (1969a, b) and Bosman and van Daal (1970); for the analogous situation of a magnetic moment in systems such as **CuMn** showing a Kondo effect, see Chapter 3, Section 8.

Since polarons are heavy, a gas of polarons in a conduction or valence band of a transitional-metal oxide can be non-degenerate at quite moderate temperatures even if the concentration is high. In this book we give examples of both degenerate and non-degenerate gases, particularly in oxides of mixed valence. We have to emphasize that the polaron concept is only valid if the number of carriers

(n) is less and perhaps considerably less than the number of sites (N), so that each carrier has a sufficient number of sites round it to "polarize". The non-degenerate gas should have the following properties. The thermopower α is given by the Heikes formula (Heikes and Ure 1961)

$$\alpha = \frac{k_B}{e} \ln\left(\frac{1-c}{c}\right),$$ (15)

where $c = n/N$. The entropy is

$$S = Nkc\left[\ln\left(\frac{1}{c}\right) - 1\right],$$ (16)

and the specific heat is zero. If the gas is degenerate then the normal formulae for the specific heat and thermopower for a metal should apply, but with the enhanced effective mass.

3 Bipolarons†

Self-trapped electrons or holes repel each other on account of their charge; on the other hand, attractive forces exist—first, because a homopolar force may be set up by the overlapping of the electron clouds, and, secondly, because the energy gained by polarizing the medium depends on the square of the field. This applies both for dielectric and spin polarons (Chapter 9). For dielectric bipolarons the concept was introduced by Lakkis *et al.* (1976) and Schlenker and Marezio (1980) to account for the behaviour of the compound Ti_4O_7. If the oxygens are supposed to be in the state O^{2-} then the chemical composition demands that here there are an equal number of Ti^{3+} and Ti^{4+} ions. At low temperatures the substance is diamagnetic; the structure is made up of pairs of Ti^{3+} ions, presumably bound together by homopolar bonds. Between about 130 and 150 K a phase exists in which the substance remains diamagnetic, but the conductivity is high, in the range $1-10^2\,\Omega^{-1}\,cm^{-1}$, and only slightly activated. The transition is of Verwey type, as for Fe_3O_4 (Chapter 8), but diamagnetic *bipolarons* are formed. There is another transition at 150 K at which the conductivity increases again and the material becomes paramagnetic. It is thought that polarons dissociate into single polarons, probably "heavy polarons" of the Eagles type, since the conductivity, of order $10^3\,\Omega^{-1}\,cm^{-1}$, is almost independent of temperature. We suggest that a non-degenerate gas of heavy particles is responsible, the resistivity being due to electron–electron collisions.

Another material, cited by Schlenker (1985), for which bipolarons probably exist is $Na_{0.3}V_2O_5$. Clear evidence is also given by Klipstein *et al.* (1985) for non-magnetic charge carriers in polyacetylene.

---

† A review of the evidence for bipolarons is given by Schlenker (1985).

4 The Peierls transition

Peierls (1955, p. 108) first pointed out that a rigid linear chain of one-electron atoms, with an interatomic distance a such that metallic behaviour is expected, is inherently unstable. For such a material the wavenumber k is $1/2a$; if each atom is moved a distance $\pm\delta$ in alternating opposite directions, either along the line or at a given angle to it, then a superlattice is set up that will produce a gap of width $V_0\,\delta/a$, where V_0 is an energy depending on the electron–phonon interaction. This depresses the energies of all electrons in the band, though the depression is greatest for those near the Fermi energy.

The energy of a state originally at $E_F - \eta$ becomes

$$E = E_F - (\eta^2 + \Delta^2)^{1/2},$$

where 2Δ is the energy gap. If $\eta \gg \Delta$, this approximates to $\Delta^2/2\eta$. If one assumes that the density of states is constant then integration over all energies gives

$$-\tfrac{1}{2}N(E_F)\Delta^2\left[\ln\left(\frac{2H}{\Delta}\right) + \tfrac{1}{2}\right]. \tag{17}$$

where H is an effective maximum for η (of order of E_F for a half-filled band). The logarithmic term in (17) ensures that the elastic term, varying as Δ^2, is such that the sum of the two has a minimum.

As the temperature rises, a second-order phase transition occurs at a temperature T_P given by

$$1.76 k_B T_P = \Delta(0),$$

$\Delta(0)$ being the gap for zero temperature (Gill 1986). An example is given in Fig. 2.6, for the blue bronze $K_{0.3}MoO_3$, showing current perpendicular to and along the b-axis discussed by Fogle and Perlstein (1972). The sign of the Hall coefficient changes at the transition, being, as we should expect, much greater in the semiconducting region.

The Peierls distortion described here is called "coherent". If, however, the band is not half-filled, but has a wave function ψ such that ka is an irrational number, then a wave-like distortion can be set up in the material with the same wavenumber k, which produces a gap, but k is incommensurate with a. Fröhlich (1950) sought to explain superconductivity through the existence of such waves, supposing them to be mobile. Although they are not related to superconductivity, they do appear to be mobile. The report by Gill (1986) summarizes the position at that date; we have not reviewed it here. Peierls distortions in three dimensions depend on an appropriate form of the Fermi surface, such that a large part of its area is flat so that the energy values there can be split by the distortion. Here, too, a transition to metallic behaviour can occur with rising temperature.

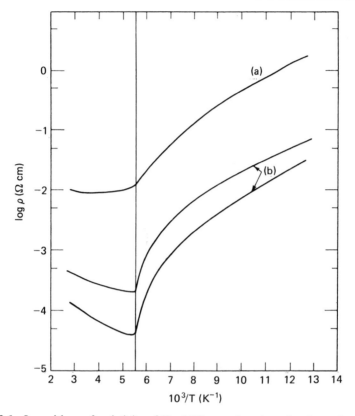

Fig. 2.6 Logarithms of resistivity of $K_{0.3}MO_3$ as a function of reciprocal temperature: (a) perpendicular and (b) parallel to the b-axis (Bouchard *et al.* 1967).

5 The Fermi energy and the Fermi surface

We turn now to the interaction energy e^2/r_{12} between electrons and consider first its effect on the Fermi surface. The theory outlined until this point has been based on the Hartree–Fock approximation in which each electron moves in the average field of all the other electrons. A striking feature of this theory is that all states are full up to a limiting value of the energy denoted by E_F and called the Fermi energy. This is true for non-crystalline as well as for crystalline solids; for the latter, in addition, occupied states in k-space are separated from unoccupied states by the "Fermi surface". Both of these features of the simple model, in which the interaction between electrons is neglected, are exact properties of the many-electron wave function; the Fermi surface is a real physical quantity, which can be determined experimentally in several ways.

The first direct experimental evidence for a sharp Fermi energy in metals was obtained from measurements of the X-ray emission by O'Bryan and Skinner

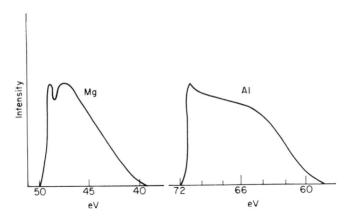

Fig. 2.7 Intensity of L_{III} X-ray emission in Mg and Al plotted against energy (O'Bryan and Skinner 1934).

(1934). Their results for the emitted intensity for Al and Mg are shown in Fig. 2.7. It will be observed that, while the limiting Fermi energy is sharp, the intensity at the bottom of the band does not follow the predicted form for the density of states, for which $N(E) = \text{const} \times E^{1/2}$. Jones et al (1934) interpreted this in the following way. Figure 2.8 represents possible transitions, shown by the dotted line. After an X-ray is emitted, a "hole" is left in the occupied states, marked ○ in Fig. 2.8. The lifetime of this state is determined by Auger transitions in which an electron A falls into the hole, giving up its energy to another electron, which is ejected to a state B above the Fermi level. The transition probability is determined by the interaction e^2/r_{12} between the electrons, and the reciprocal of the time before such a transition occurs also by the number of electrons capable of making the transition. But this number tends to zero as the energy of the hole nears that of the Fermi level.† Thus the lifetime of a hole just below the Fermi level is very large. This is why the Fermi level is sharp but the lower limit broad.

The argument due to Landau (1957) that in a perfect crystal at zero temperature the Fermi *surface* is a sharply defined physical quantity is similar. A hole just below the Fermi surface, or an electron just above it, will have their energies as functions of k somewhat affected by interaction with all the other electrons (Nozières and De Dominicis 1969); in fact, round each electron there will be a region of low charge density from which the other electrons keep away. We refer to the electron (or hole) with its surrounding region as a "pseudoparticle". But again, the lifetime of the pseudoparticle approaches infinity as its energy approaches the limiting Fermi energy. The wave vector **k** therefore remains a good quantum number for such pseudoparticles, and a Fermi surface can be defined.

---

† If ΔE is the energy separation then the lifetime is proportional to $(\Delta E)^2$ (Ziman 1964a, p. 415).

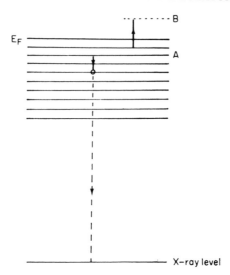

Fig. 2.8 Mechanism of Auger broadening in X-ray emission.

An important property of the Fermi surface is that the volume (in k-space) that it encloses is not altered by the interaction between the electrons, unless long-range antiferromagnetic order is set up. This was first shown by Luttinger (1960). We shall make use of this theorem in Chapter 4, Section 3 in discussing metal–insulator transitions due to correlation.

In an imperfect crystal or amorphous material the wavenumber k is not a good quantum number, and if $\Delta k/k$ becomes comparable to unity then the concept of a Fermi surface has little meaning. Nevertheless, at zero temperature a sharp Fermi *energy* must still exist.

6 Electron–electron collisions (Landau–Baber scattering)

Another effect that depends on interaction between electrons is the term in the electrical resistivity proportional to T^2 first predicted by Baber (1937) and Landau and Pomeranchuk (1937). Baber's argument was as follows. An electron in a state near the Fermi energy E_F can transfer momentum to another electron only if its energy is within $\sim k_B T$ of E_F; otherwise one electron or the other will finish up in an occupied state. This means that only a fraction $k_B T/E_F$ of all possible collisions can scatter an electron at the Fermi surface. But again the collisions must be limited to those in which the energy exchange between the particles is limited to $\sim k_B T$; this introduces another factor $k_B T/E_F$. So the mean free path l is of order given by

$$l^{-1} \sim NA\left(\frac{k_B T}{E_F}\right)^2,$$

where N is the number of electrons per unit volume and A the collision cross-section between two electrons. We thus have for the resistivity

$$\rho \sim \frac{m v_F}{N e^2 l} = \frac{m v_F A}{e^2} \left(\frac{k_B T}{E_F} \right)^2. \tag{18}$$

Another way of showing that ρ behaves as T^2 is to note that the rate of decay due to the Auger effect of an electron with energy ε_1 above the Fermi energy is (Ziman 1964a, p. 415) proportional to $(\pi k_B T)^2 + \varepsilon_1^2$; for the small values of ε_1 produced by a field, this gives a time of relaxation proportional to T^2 (Hodges et al. 1971); these authors find a proportionality with $T^2 |\ln (k_B T/E_F)|$.

Two points may be noted about Baber scattering.

(a) The resistivity plotted as a function of T should flatten out when the gas becomes non-degenerate ($k_B T \sim E_F$), as shown schematically in Fig. 2.9.

(b) For a spherical Fermi surface, current is conserved in electron–electron collisions and there is no effect on the resistance, though there is on the thermopower. Electron–electron interaction does not affect this result, as long as it is not strong enough to set up an antiferromagnetic superlattice (Monecke 1972). But umklapp scattering will, and, according to calculations reported by Bass (1972, particularly p. 477), the term due to this effect should be as large as for materials with a complex Fermi surface.

The earliest experiment observing T^2 behaviour was that of de Haas and de Boer (1934) on platinum (before the theory was presented), and Baber's paper suggested that the effect should be large for transitional metals, following Mott's (1935, 1936) model of these materials. This model proposed that parts of the Fermi surface were (s–p)-like, with small effective mass, and that these carried the current; collisions were assumed to be mainly into d-like parts of the surface, because of the assumed high density of states there, and for Baber scattering most collisions were with d-like electrons, again because of the high density of states, responsible for the high electronic specific heat. This explanation was queried by Ruthruf et al. (1978), but calculations by Potter and Morgan (1979), using realistic forms of the Fermi surface, showed that the large values of the T^2 term could be explained.

The arguments are reviewed by Kaveh and Wiser (1984). They also describe observations of T^2 behaviour in alkali and noble metals and—a point relevant to this book—the application to non-crystalline materials and other materials with short mean free path. They show that the time of relaxation τ resulting from electron–electron scattering contains an additional term and is of the form

$$\frac{1}{\tau} = A T^2 + B T^{3/2},$$

B becoming significant only when the mean free path resulting from disorder is short.

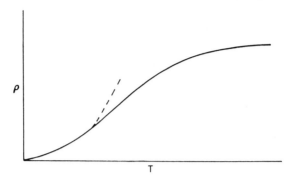

Fig. 2.9   The term in the resistivity due to Baber scattering; the dashed line shows the T^2 term.

However, for the resistivity of non-crystalline materials at low temperatures, the most important effect is on L_i, the inelastic diffusion length. We have seen (Chapter 1, Section 10) that this is given by $L_i = (D\tau_i)^{1/2}$ and that there is a term in the conductivity proportional to l/L_i, so that electron–electron scattering gives a negative term in the resistivity proportional to T^ν, with ν between 1 and $\frac{3}{4}$.

A term in the resistivity proportional to T^2 is not necessarily due to Baber scattering. Thus in ferromagnetic materials scattering by magnons produces a term of this kind (Mott 1964), and it is believed to be a fairly common consequence of magnetic excitations (see Chapter 3). In Chapter 6 we discuss the behaviour of "metallic" transitional-metal oxides near a metal–insulator transition, where the strong mass enhancement due to correlation should give a large T^2 term, though a description in terms of magnetic excitations may be equally valid.

7   Excitons

A well-known phenomenon that does not exist in a model of non-interacting electrons is the "exciton". In a non-conducting crystal the only electronic excitations possible if one neglects electron–electron interaction are the production of a free electron and a free hole, so that the absorption of radiation should begin when the absorbed energy $h\nu$ is equal to the band gap. Frenkel (1931) and Peierls (1932) were the first to point out that the optical absorption spectrum of an ideal insulating crystal in which the atoms are supposed fixed in position would consist of a series of lines leading up to a series limit. This is because electrons and holes attract each other, their mutual potential energy being $-e^2/\kappa_\infty r$, where κ_∞ is the high-frequency dielectric constant. These will form a series of bound states with energies

$$-\frac{m^* e^4}{2\hbar^2 \kappa_\infty^2} \frac{1}{n^2}, \tag{19}$$

where $n = 1$ corresponds to the lowest state of the exciton and where

$$\frac{1}{m^*} = \frac{1}{m_1} + \frac{1}{m_2}.$$

m_1 and m_2 are the effective masses of electron and hole. n has the integral values 1, 2, 3, ... The complex of electron and hole can move with wavenumber k, and as such is called an "exciton". This description of an exciton is due to Wannier (1937) and Mott (1938) and is always appropriate to highly excited states (see Mott and Gurney 1940, p. 90).

The earlier formulation due to Frenkel (1931) is appropriate to molecular crystals where the overlap between the orbitals even of excited states is weak. If $\phi_i(q)$ is the wave function for a molecule in its ground state and $\phi'_i(q)$ that in an excited state, then a Slater determinant Ψ_s formed from the product

$$\phi_1(q_1)\phi_2(q_2)\ldots\phi'_s(q_s)\ldots\phi_n(q_n)$$

represents the wave function when molecule s is excited. This is not an eigenstate of the system; the eigenstates are

$$\sum_s e^{i\mathbf{k}\cdot\mathbf{a}_s}\Psi_s, \tag{20}$$

which represent the Frenkel exciton moving with wavenumber $\mathbf{k}$.

Thus in either formulation the exciton spectrum consists of a series of bands, but the optical absorption spectrum consists of a series of lines because the selection rule

$$\mathbf{k} + \mathbf{q} = 0,$$

where $\mathbf{q}$ is the wavenumber of the light, picks out a definite value of $\mathbf{k}$.

There is a wide literature on excitons: an early review is that by Knox (1963), and a book dealing exhaustively with the subject is Rashba and Sturge (1982). The aspects most relevant to the discussions in the present book are

(a) excitonic effects near the metal–insulator transition; electrons and holes present in small numbers may in certain circumstances combine or crystallize to form a non-conducting "excitonic" state (see Chapter 4);

(b) the possibility that excitons may exist in metals, which we discuss in the next section.

8   Excitons in metals

The problem of whether optical absorption in metals can show exciton lines is closely related to the considerations of the last section, and a short discussion will be given here. Experimentally, the addition of free carriers to a non-metal leads first to the broadening and eventually the disappearance of exciton lines. Figure 2.10 shows the results of Wilson and Yoffe (1969) on WSe_2 doped with $NbSe_2$, the Nb atom containing one fewer d-electron than the W atom. Somewhat similar results on Mg–Bi are due to Slowik and Brown (1972).

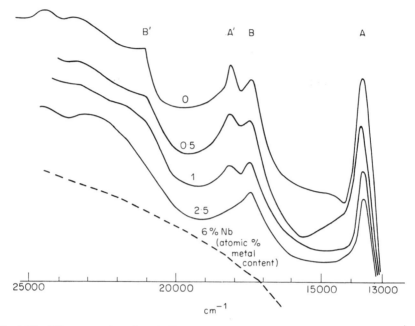

Fig. 2.10   Effect on exciton lines in $W_{1-x}Nb_xSe_2$ of Nb content (Wilson and Yoffe 1969).

From the theoretical point of view, there are two ways in which free carriers can affect exciton formation.

(a) They can screen the Coulomb field between electron and hole, so that no bound state can be formed.

(b) They can shorten the lifetime of the exciton and hence broaden the absorption line by allowing Auger transitions, in which electrons and holes recombine by giving their energy to free carriers. This can occur whether the carriers are free, localized in the Anderson sense by disorder or bound to centres.

We shall treat the second effect first, limiting ourselves to the case when the hole has a high effective mass and can be regarded as stationary as in X-ray absorption. The problem, first treated by Mahan (1967a, b), is then related to the peaks due to many-body effects at the X-ray absorption and emission edges, predicted by Nozières and De Dominicis (1969), Friedel (1969) and Hopfield (1969) and now widely observed. Figure 2.11 shows peaks in the $L_{II, III}$ emission spectrum of magnesium and of aluminium observed by Senemaud and Hague (1971). The peak in Mg was observed first by O'Bryan and Skinner (1934) and thought to be due to overlapping bands for a divalent metal, a conclusion that was reinforced by its absence in K emission. In Fig. 2.12, however, we show the results due to Neddermeyer (1971) for the Mg–Al system; the peak persists with varying composition, which is not compatible with a band-structure explanation.

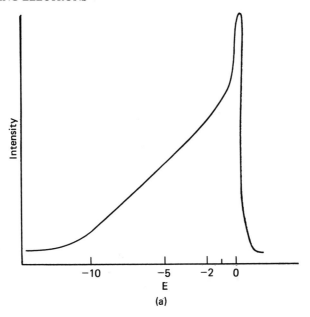

(a)

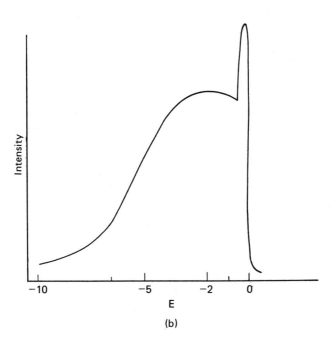

(b)

Fig. 2.11 $L_{II,III}$ emission bands (Senemaud and Hague 1971): (a) Al; (b) Mg.

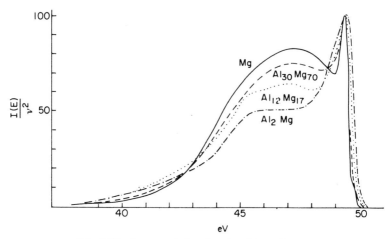

Fig. 2.12 $L_{\mathrm{II, III}}$ emission bands of Mg–Al alloys.

The explanation of this peak is as follows. Suppose that the number of conduction electrons is small, so that the Coulomb field is not screened out and that a hole in the X-ray level creates an exciton level below the bottom of the conduction band. The levels are shown in Fig. 2.13. Then an exciton absorption line should be possible. But the sudden change in field will produce excitations of electrons at the Fermi level, so that the exciton line is broadened as shown in Fig. 2.14(a). Also, we do not expect a sharp increase in absorption when the electron jumps to the Fermi level, leaving the exciton level A in Fig. 2.13 unoccupied, because of the very large Auger broadening due to transitions from the Fermi level into this unoccupied state.

As the number of electrons in the conduction band increases, the situation goes over continuously to that described by Nozières and De Dominicis and illustrated in Fig. 2.14(b). The peak is obtained as follows. At the moment in time when the transition occurs, the wave functions $\psi_i(x_i)$ of all the other conduction electrons have to change to new functions $\psi_i'(x_i)$, which screen the positive charge left in the inner level. So the transition probability must be multiplied by the product

$$\prod_i \int \psi_i^* \psi_i' \mathrm{d}^3 x.$$

This in general gives a logarithmic divergence, as shown by Noziéres and De Dominicis, so the transition probability is infinite. The exciton state, if one exists, is of course filled, and this model provides one description of the way the hole is screened (see Friedel 1952a, b).

The question of whether the carriers screen the hole and so prevent the formation of any exciton level has not been discussed recently (see Cauchois and Mott 1949). Clearly this is a *qualitative* rather than an exact question, for the

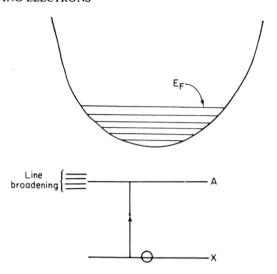

Fig. 2.13 Formation of excitons in a metal.

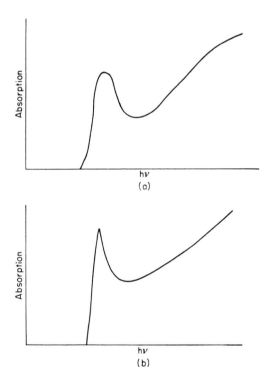

Fig. 2.14 Expected absorption spectrum: (a) for a small number of free electrons; (b) for a normal metal.

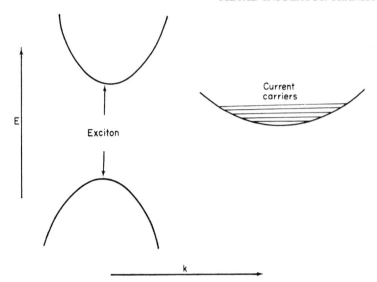

Fig. 2.15   Suggested band structure for WSe$_3$ doped with Nb.

reasons given above. It seems likely that the screening is sufficient to prevent a bound state, because the empirically successful condition $(n^{1/3}a_H > 0.25)$ for metallic conduction (Chapter 4) is just the same as that for the non-existence of the exciton. Wilson and Yoffe (1969) point out, however, that in WSe$_3$ doped with Nb the carriers are not in the same band as the electrons bound in the exciton, the situation being as in Fig. 2.15, so the values of a_H for the two bands should differ, and non-metallic behaviour and exciton formation may occur for different concentrations. The measurements of Raz *et al.* (1972) may be of relevance here; they investigated the disappearance of Wannier excitons in solid Xe–Hg films at 10–40 K. Their results are shown in Fig. 2.16.

9   The Hubbard intra-atomic energy

Much of this book is concerned with the properties of narrow bands to which the tight-binding approximation is appropriate. In this case, if the band is half full or nearly so, the short-range repulsion between the electrons may have very important effects on the properties of the electrons in the bands, producing magnetic moments and non-conducting properties. These are a major theme of this book. At this point we introduce the Hubbard intra-atomic energy†

$$U = \left\langle \frac{e^2}{r_{12}} \right\rangle,$$

---

† For donors in semiconductors e^2/r_{12} should be replaced by $e^2/\kappa r_{12}$, where κ is the dielectric constant.

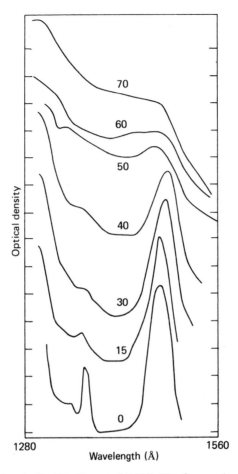

Fig. 2.16 Exciton lines in Xe–Hg (Raz *et al.* 1972). The figures show the mole fractions of Hg.

where $\langle\ \rangle$ denotes an average over a single site, so that we may write

$$U = \int \int \frac{e^2}{r_{12}} |\phi(r_1)|^2 |\phi(r_2)|^2 \, d^3x_1 \, d^3x_2, \tag{19}$$

where $\phi(r)$ is the wave function of an electron on the site. If $U \gtrsim B$, where $B = 2zI$ is the bandwidth, then the approximations of this chapter may break down completely. Even if this not so, the effect of U may be important, as we shall see in subsequent chapters, particularly Chapters 4 and 6.

For hydrogen wave functions $\phi = (\alpha_0^3/\pi)^{1/2} e^{-\alpha_0 r}$ the value of U has been evaluated and is (Schiff 1968, p. 258)

$$U = \tfrac{5}{8}\alpha_0 e^2. \tag{20}$$

More generally, for many-electron atoms we define U as the energy required to transfer an electron from one atom to another, so that

$$U = (\text{ionization energy}) - (\text{electron affinity}).$$

For an ion such as Mn^{2+} in the state $3d^5$, where the electron must be transferred to a state with opposite spin, clearly

$$U = U_0 + 4J_H,$$

where J_H is the Hund's-rule coupling (Chapter 3, Section 1). If two possibilities exist, for example for a state $3d^3$, the value would be $U_0 + 2J_H$ for transference to an antiparallel state and $U_0 - J_H$ for transference to a state with antisymmetrical orbital wave function for all the electrons.

10   Effect of the Hubbard U on Anderson localization

We discuss in this section the effect of short-range interaction on the Anderson-localized states of a Fermi glass described in Chapter 1, Section 7, and in particular the question of whether the states are singly or doubly occupied. Ball (1971) was the first to discuss this problem. In this section we consider an electron gas that is far on the metal side of the Wigner transition (Chapter 8); the opposite situation is described in Chapter 6, where correlation gives rise to a metal–insulator transition. We also suppose that Anderson localization is weak ($\alpha a \ll 1$); otherwise it is probable that all states are singly occupied.

Our argument follows that of Mott (1972a) and is illustrated in Fig. 2.17. We suppose that if two electrons are in the same occupied state then the mean of their interaction energy $\langle e^2/r_{12} \rangle$ averaged over the whole state can be written as ΔE. Then these states will be singly occupied down to an energy ΔE below the Fermi energy. An energy

$$E_M = E_F - \Delta E$$

separates singly from doubly occupied states, as shown in Fig. 2.17, and the number n_1 of singly occupied states is

$$n_1 = \int_{E_F - \Delta E}^{E_F} N(E)\, dE \approx \Delta E\, N(E_F). \tag{21}$$

Our problem is to estimate ΔE. It is the mean repulsive energy of a pair of charges at a distance α^{-1} from each other. This will depend on the effective dielectric constant of the electron gas. This should be large for weak Anderson localization and will effectively screen out the repulsion, *except* when both electrons are in the same atom. We therefore write

$$\Delta E \approx \tfrac{4}{3}\pi(\alpha a)^3 U, \tag{22}$$

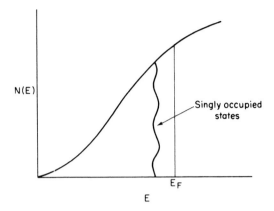

Fig. 2.17 Density of states and singly and doubly occupied states for a Fermi glass (Mott 1972a).

where, as in the last section, U is the Hubbard intra-atomic energy for a single atom. From (21) and (22) it follows that the number of singly occupied states is

$$UN(E)(\alpha a)^3,$$

so that the mean distance between them is

$$\{[UN(E)]^{1/2}\alpha a\}^{-1},$$

which will be of the same order as α^{-1}. Thus the overlap between the singly occupied states is strong, and the moments should be strongly coupled.[†] Correlation should therefore introduce weak moments, strongly coupled, fluctuating over a distance α^{-1}; the moment in each fluctuation is about a Bohr magneton μ, or on each atom $\mu(a\alpha)^3$.

The term U will raise the doubly occupied states in a many-electron model, so that at the Fermi energy there should be a mixture of singly and doubly occupied states. Below ΔE all are doubly occupied.

---

† Kaplan *et al.* (1971) have also discussed this problem. They also propose that states can be singly or doubly occupied, but come to a different conclusion about the coupling between the moments.

3

Magnetic Moments

1 Introduction

A theory of non-interacting electrons fails to predict any magnetic moments in solids. In this chapter we discuss certain magnetic phenomena, all of which depend on the intra-atomic repulsion, i.e. the Hubbard U, which is important for the consideration of one kind of metal–insulator transition (Chapters 4 and 6). These are antiferromagnetism, spin polarons and the conduction band in antiferromagnetic insulators, magnetic moments in impurities and the Kondo effect. In Chapter 6 we also discuss impurity bands in antiferromagnetic insulators, a problem relevant to high-temperature superconductors (Chapter 9).

We first note that an isolated atom with an odd number of electrons will necessarily have a magnetic moment. In this book we discuss mainly moments on impurity centres (donors) in semiconductors, which carry one electron, and also the d-shells of transitional-metal ions in compounds, which often carry several. In the latter case coupling by Hund's rule will line up all the spins parallel to one another, unless prevented from doing so by crystal-field splitting. Hund's-rule coupling arises because, if a pair of electrons in different orbital states have an antisymmetrical orbital wave function, this wave function vanishes where $r_{12} = 0$ and so the positive contribution to the energy from the term e^2/r_{12} is less than for the symmetrical state. The antisymmetrical orbital state implies a symmetrical spin state, and thus parallel spins and a spin triplet. The two-electron orbital functions of electrons in states with one-electron wave functions $a(x)$ and $b(x)$ are, to first order,

$$2^{-1/2}[a(1)b(2) \pm a(2)b(1)],$$

and so the difference between the energies of these terms is $2J_H$, where

$$J_H = \int \int a(1)b(2)Ha(2)b(1) \, d^3x_1 \, d^3x_2 \qquad (1)$$

is the exchange integral, H being the Hamiltonian; J_H is normally of order 1 eV.

2 Antiferromagnetism

In this section, in discussing a typical antiferromagnetic crystal like MnO, we suppose as indicated in the last section that each Mn^{2+} ion has a moment due to its five electrons lined up by Hund's rule, and that the crystal is an insulator.

The arrangement of moments in a typical antiferromagnetic is shown in Fig. 3.1. Moments on adjacent atoms are, in the simplest cases such as MnO or NiO, coupled so that they are antiparallel. Then at low temperatures the susceptibility depends on whether the field is parallel or perpendicular to the moments. If it is parallel, the susceptibility $\chi_\parallel$ goes to zero as $T \to 0$. This is because the coupling prevents any spins from turning over, and no spin waves are excited. On the other hand, if the field is perpendicular to the moments, they will be oriented by the field as shown in Fig. 3.2 and $\chi_\perp$ is independent of temperature. The susceptibilities in the two directions are shown in Fig. 3.3. In practice, a macroscopic sample will usually contain numerous domains, and then the average susceptibility is

$$\chi = \tfrac{1}{3}\chi_\parallel + \tfrac{2}{3}\chi_\perp.$$

Above a certain temperature T_N, known as the Néel temperature, the long-range ordering of the moments disappears and the susceptibility then behaves roughly as

$$\chi = \frac{N\mu^2}{k_B(T+\Theta)}, \tag{2}$$

where $-\Theta$ is called the (negative) paramagnetic Curie temperature.

The ground state of an isotropic antiferromagnet is a singlet, the moments shown in Fig. 3.1 rotating in much the same way as the moment on an impurity rotates in the Kondo problem (see Section 8). The time for rotation, however, is very long, according to Anderson (1952) of order

$$hn/k_B T_N,$$

where n is the number of atoms in the specimen and T_N the Néel temperature. For specimens of reasonable size this time is very large, perhaps years, and so effects due to this rotation are negligible.

In a crystalline antiferromagnet the moment on each ion is less than it would be on the free ion. There are two separate phenomena involved here. One is the zero-point energy of the spin waves, which reduces the moment on each ion by a factor (Anderson 1952, Ziman 1952)

$$1 - \frac{1}{2z},$$

where z is the coordination number. This is not important for our discussions in this book. The second is due to the same cause as the interaction that leads to antiferromagnetism. We suppose that I is the overlap energy integral between nearest-neighbour atoms already introduced in Chapter 1, equation (10), and U is

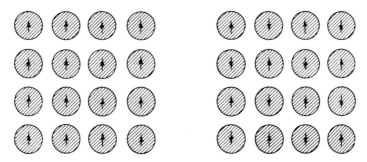

Fig. 3.1 Arrangement of moments in a typical ferromagnetic and antiferromagnetic crystal (Ziman 1964a).

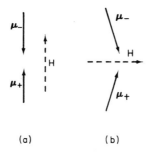

(a)   (b)

Fig. 3.2 Orientation of moments in an antiferromagnatic material when field is (a) parallel, (b) perpendicular.

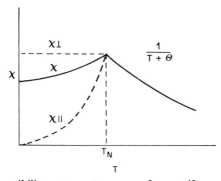

Fig. 3.3 Typical susceptibility–temperature curves for a antiferromagnetic insulator with field parallel and perpendicular to moments.

the (intra-atomic) repulsion between a pair of electrons on the same atom. For 1s-states this is given by

$$U = \tfrac{5}{8}e^2/a_{\mathrm{H}}, \tag{3}$$

where a_{H} is the hydrogen radius. For 3d-states in nickel it has been estimated by Watson (1960) to be 20 eV. As we shall see in Chapter 6, a much smaller screened value U_{scr} is appropriate in considering the metal–insulator transition in transitional-metal oxides; the unscreened value given here is appropriate for estimating the magnetic coupling.

To evaluate both the coupling and the magnitude of the moments, we consider first an array of one-electron atoms in non-degenerate states, and let ϕ_i denote an atomic orbital for an electron on atom (i). Then the electron on atom (i) can overlap on to neighbouring atoms (j), so we may write its wave function in the form

$$\psi_i = \phi_i + A\textstyle\sum \phi_j, \tag{4}$$

where the sites (j) are the nearest neighbours to (i) and the summation is over the z nearest neighbours to atom (i). A is a parameter that is to be determined. The energy due to the term $A\sum\phi_j$ in the antiferromagnetic state is

$$z(2AI + A^2U). \tag{5}$$

Choosing A so that (5) has its minimum value, we find

$$A = -I/U,$$

so the moment on each atom is reduced by the factor

$$1 - zI^2/U^2, \tag{6}$$

and the energy difference between the antiferromagnetic and ferromagnetic arrangements is

$$zI^2/U \tag{7}$$

(see Anderson 1963). Half this would be the energy needed to disorder the spins. As a rough estimate of the Néel temperature T_{N}, we equate, for $s = \tfrac{1}{2}$, $k_{\mathrm{B}}T_{\mathrm{N}}\ln 2$ to $\tfrac{1}{2}zI^2/U$, so that

$$k_{\mathrm{B}}T_{\mathrm{N}} = \frac{zI^2}{2U\ln 2} = \frac{B^2}{4zU\ln 2},$$

where $B = 2zI$ is the bandwidth.

For a transition of "Mott" type we shall show in Chapter 4, Section 3, neglecting the discontinuity resulting from long-range forces, that the transition should occur when $2zI = U$. Near the transition the energy needed to excite an electron into the upper Hubbard band is $U - 2zI$. The wave function of an electron then falls off as $e^{-\alpha r}$, where $\alpha = 2m(U - 2zI)^{1/2}/\hbar^2$. Thus the amount of spin in the sphere surrounding each atom will be made up from electrons on many of the surrounding atoms, and will clearly go to zero as $\alpha \to 0$. Equation (6) is thus an underestimate for the reduction in the moment.

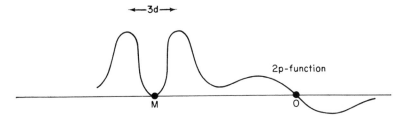

Fig. 3.4   Wave functions of 3d-electrons overlapping through 2p-orbitals on oxygen ions.

In doped semiconductors I is due to direct overlap; in transitional-metal oxides the overlap between the d-orbitals is frequently *via* the oxygen ions and is then often called a superexchange interaction. Figure 3.4 shows the kinds of wave function expected. As regards magnitudes, if $B \sim 1$ eV, $U \sim 10$ eV and $z = 4$, $k_B T_N$ should be 0.01 eV so that $T_N \sim 100$ K, which shows why low Néel temperatures are common.

3   Antiferromagnetism and ferromagnetism in transitional-metal compounds

In any cubic field, or in the case of octahedral coordination by eight anions, the 3d-state of a transitional-metal cation splits into six states of t_{2g} symmetry with orbital wave functions of the form

$$xyf(r), \quad yzf(r), \quad zxf(r)$$

and four states with e_g symmetry and wave functions

$$(x^2 - y^2)f(r), \quad (z^2 - x^2)f(r).$$

The energy difference between these states, or states split by fields of other symmetries, is called the "crystal-field splitting". In considering the spin of the transitional-metal ion, we often have to ask whether the crystal-field splitting or Hund's-rule coupling is the larger. If the crystal-field splitting is larger then the ion may be forced into a "low-spin" state. Thus an Fe^{2+} ion in the $3d^6$ state may be forced into a state with zero spin only in the e_g band. A review of low-spin materials is given by Wilson (1972, p. 168). They include FeS_2, in which Fe in the d^6 state in the t_{2g} band has *no* spin, so this material is a non-magnetic non-metal. On the other hand, if the Hund's-rule term is the larger, Co^{2+} will have a spin moment of 3 Bohr magnetons. Transitions from high to low spin can occur under pressure, for instance for Fe ions replacing Mn in MnS_2, where the transition occurs between 40 and 130 kbar (Bargeron *et al.* 1971).

Antiferromagnetic coupling will occur in oxides such as MnO, where there are five collinear spins on each $Mn^{2+}(3d^5)$ ion, and the overlap from neighbouring ions is into states with opposite spin. On the other hand, ferromagnetic coupling can also occur if large overlap is possible into unoccupied states with the same spin as that on the ion under consideration. An example is

$CrCl_3$, which is ferromagnetic. The three electrons in the $Cr^{3+}(3d^3)$ ion are coupled by Hund's rule and just fill the t_{2g} state for a given spin. The overlap into the higher e_g states must give stronger Hund's-rule coupling than overlap into the t_{2g} states with opposite spin. In another ferromagnetic oxide, EuO, the overlap must be into 5d-states of the Eu ion.

An appropriate analysis is as follows. We set instead of (4) for the orbital on atom i

$$\phi_i + B\sum \phi'_j,$$

where ϕ_i is the t_{2g} orbital on atom i and ϕ'_j is the e_g orbital (for the case of $CrCl_3$) on neighbouring sites j. If ΔE is the energy interval between t_{2g} and e_g states and J_H the Hund's-rule coupling energy between electrons in t_{2g} and e_g states on the same atom then, instead of (5), we should write

$$z[2BI' + B^2(U + \Delta E - J_H)], \tag{8}$$

where I' is the energy overlap integral between e_g and t_{2g} states on neighbouring atoms. If the atoms are collinear, this will vanish, and ferromagnetism can only arise if they are not. According to Goodenough (1963, p. 184), for d^3 coupling via anions the change from antiferromagnetic to ferromagnetic coupling occurs at angles between 125° and 150°. The energy resulting from (8) is

$$-\frac{zI'^2}{U + \Delta E - J_H}. \tag{9}$$

For ferromagnetism to occur, this must be *lower* than (7).

For later discussions it is important to consider what happens if one of the d-orbitals (e_g or t_{2g}) is not full (e.g. CoO with seven electrons per atom or $TiCl_3$ with one). The d-orbital is degenerate, and the orbital motion is no longer quenched. Consider the case of a transitional-metal ion in the d^1 state. Suppose that the spins are parallel to the z-axis; the spin–orbit coupling will separate the orbitals into functions of the forms

$$f(r)\sin 2\theta\, e^{i\phi},$$
$$f(r)\sin 2\theta\, e^{-i\phi},$$
$$f(r)\cos 2\theta.$$

It is not, however, sufficient to take the first two functions on alternate atoms; the addition of a term of the type $Af(r)\cos 2\theta$, giving an atomic orbital of the form

$$f(r)(\sin 2\theta\, e^{\pm i\phi} + A\cos 2\theta), \tag{10}$$

will always increase the antiferromagnetic coupling. This will lead to a canting of the spins (Mott and Zinamon 1970). Such canting has in fact been observed in CoO by Kahn and Erickson (1970). In Chapter 4, Section 7 we argue that the canting for antiferromagnetic insulators must be such as to split the d-*band* into full and empty sub-bands.

Also of interest is the case when the orbital angular momentum is quenched by the ligand fields, as for d^4 or d^9 cations in octahedral interstices, or d^1 or d^6 cations in tetrahedral interstices (Goodenough 1963, p. 66). There is then a single occupied e_g orbital, and combining this with another e_g orbital *cannot produce an orbital momentum*. There can thus be no spin–orbit coupling. In such a situation for an isolated moment a distortion of the lattice, known as the Jahn–Teller effect, will always occur in such a way as to remove the orbital degeneracy. In Chapter 6, Section 2 we mention this as the reason why in MnO the carrier Mn^{3+} (in the state d^4) shows polaron hopping; the "polaron" is a Jahn–Teller distortion. In Fe_3O_4 the absence of Jahn–Teller distortion is discussed in Chapter 8, Section 2, where we suggest that the splitting shown by Mössbauer measurements is due to a removal of degeneracy by canting as proposed above.

In simple cubic compounds like NiO, MnO and CoO the metal ions lie on a face-centred cubic. As pointed out by Ziman (1952), for this structure and for spherical orbitals, antiferromagnetism with a finite Néel temperature must be due to interaction between next-nearest neighbours, because in any antiferromagnetic structure each moment will have as many parallel as antiparallel neighbours. In NiO and CoO the orbitals are not spherical, but in MnO the $3d^5$ ion is spherical. In this compound the Néel temperature is therefore anomalously low, and there remains abnormally strong short-range order above the Néel temperature (Battles 1971).

Antiferromagnetic insulators of the kind discussed here are sometimes called "Mott insulators". It was thought until recently that a crystalline array of one-electron atoms at a sufficient distance apart would necessarily be of this type. In fact, however, many crystals in which each metal atom has a spin $\frac{1}{2}\hbar$ and that are insulating do not show an antiferromagnetic lattice or pairing of atoms; examples are TiI_3, TiOX (X = Cl, Br, or I) (Maule *et al.* 1988). They show a small positive susceptibility independent of temperature. These may possibly be the "resonating valence bond" materials discussed by Anderson (1973, 1987). The only one-electron antiferromagnet known to us is VF_4 (Gossard *et al.* 1970); there are also "one-hole" materials such as La_2CuO_4, which on doping with Sr or Ba become high-temperature superconductors (Chapter 9).

The discussion by Maule *et al.* (1988) does not mention the hypothesis of Anderson that a resonating valence bond state exists, but we think that α-TiX_3, β-TiX_3 and TiOX (X = Cl, Br, I) are probably of this type. The calculations of Callaway (1987) (for a simple hexagonal structure), indicate that this should occur, giving a singlet ground state, for $U/t > 14.69$, i.e. for large U. Maule *et al.* find that $ScBr_3$, with no d-electrons, shows an optical absorption spectrum with p→d transitions leading to excitons and a band edge at about 3.5 eV. A similar p→d absorption edge is seen in α-$TiBr_3$ and TiOCl, but both show a weaker absorption band between about 1.3 and 2 eV. This is ascribed to transitions of the d electrons from e_g to t_{2g} states. The observed magnetic susceptibility is small $(0.5-1) \times 10^{-3}$ c.g.s. units; in TiOCl it depends little on temperature; in α-$TiBr_3$ a rapid drop by a factor of about 3 at 200 K is attributed to a change of structure. The observed structures do not allow pairing, as in VO_2. Resonating valence bond states should not show any Pauli susceptibility, and the observed effect may be of Van Vleck type, though it is perhaps larger than we might expect.

There appears to be no theory, other than that of resonating valence bonds, that can account for this behaviour.

4   The conduction band of an antiferromagnetic non-metal; spin polarons

In this section we discuss the properties of an electron in the conduction band of an antiferromagnetic insulator. This may be a simple Mott insulator, but, since the experimental evidence is related to them, we first discuss materials like EuSe, where the europium ion has seven 4f-electrons and electrons can be introduced into the conduction band by doping with GdSe; the ion Gd^{2+} has the same number of f-electrons but one more electron in an outer shell, so a Gd ion acts as a donor.

The interaction between an electron spin in an s- or d-like conduction band and a given 4f-moment can be written as

$$J_{sf}sS|\psi_s(0)|^2\Omega. \tag{11}$$

Here J_{sf} is the Hund's-rule energy coupling the spin s of the conduction electron and the moment S on the europium ion in a ferromagnetic sense. Ω is the atomic volume and ψ_s is the wave function of the conduction electrons.

The excitations of the spin system are magnons. The strong scattering of conduction electrons by magnons has been treated by a number of authors (de Gennes and Friedel 1958, Brinkman and Rice 1970a). Because of this strong scattering, except near the bottom of a conduction band, k will not be a good quantum number for a conduction electron.

States near the bottom of a band are, however, of major interest in considering the properties of semiconductors. Here magnetic semiconductors differ from normal semiconductors chiefly in that the carrier can form a "spin polaron" (a concept first introduced by de Gennes 1960). That is to say, over a sphere of radius R round the carrier, it can orient the moments parallel or antiparallel to its own, and can move with its spin cloud much as a dielectric polaron can move (Chapter 2, Section 2). The following discussion was first given by Mott and Davis (1971, p. 176). Let m be the mass of the electron in the conduction band (possibly enhanced by dielectric polaron formation). Its kinetic energy in this spherical region is then $\frac{1}{2}\hbar^2\pi^2/mR^2$. The interaction J_{sf} between the carrier and the moments is independent of R. Let J_N be the energy per moment needed to go from the antiferromagnetic arrangement to the ferromagnetic. Then the energy of the polaron is

$$\frac{\hbar^2\pi^2}{2mR^2}+\frac{4\pi}{3}\left(\frac{R}{a}\right)^3 J_N-J_{sf}.$$

This is a minimum when

$$R^5=\hbar^2\pi a^3/4mJ_N, \tag{12}$$

which gives a large value for the polaron radius R (as we would expect) if J_N is small. The total energy is

$$\frac{5\hbar^2\pi^2}{6m}\left(\frac{4mJ_N}{\pi\hbar^2 a^3}\right)^{2/5} - J_{sf}, \tag{13}$$

so a spin polaron results only if this is negative.

A spin polaron should move at low temperatures with a fixed wave vector k, like any other pseudoparticle, and be scattered by phonons and magnons. The effective mass is expected to be of the form $me^{\gamma R/a}$, where $\gamma \sim 1$. To obtain this result, we compute the transfer integral when the polaron moves through one atomic distance. The spin will contribute a term proportional to

$$\prod_r \cos\theta_{r,r+1}, \tag{14}$$

where θ is the change in orientation when the carrier moves through one atomic distance. We may expect that $\theta_{r,r+1} \sim a/R$, so (14) becomes

$$\prod_n \left(1 - \frac{a^2}{2R^2}\right)^n, \tag{15}$$

where $n = (R/a)^3$, and therefore for large R will vary as

$$\text{constant} \times e^{-\gamma R/a}. \tag{16}$$

For free particles this should be $\hbar^2/m_p a$, and we see that for large R the effective mass n_p may be large.

Above the Curie or Néel point, a spin polaron will move by a diffusive process. A moment on the periphery of the polaron will reverse its direction in a time τ (the relaxation time for a spin wave). Each time it does so, the polaron can be thought to diffuse a distance $(a/R)^3 R$, so the diffusion coefficient is

$$D \approx \tfrac{1}{6}a^6/R^4\tau, \tag{17}$$

decreasing rapidly with increasing R. This should be independent of temperature.

An alternative treatment of the effective mass of an antiferromagnetic spin polaron is given by Kasuya (1970), who also obtained in a one-dimensional model a high effective mass. The theory was also discussed by Nagaev (1971).

5 Magnetic semiconductors

Perhaps the most direct evidence of the high effective mass in such materials comes from the work of Shapira et al. (1972) on the conductivity of EuTe and the work of Shapira and Reed (1972) on EuS, an antiferromagnetic material, with sufficient non-stoichiometry to make it a degenerate n-type semiconductor. Figure 3.5 shows the temperature variation of the resistivity for various magnetic

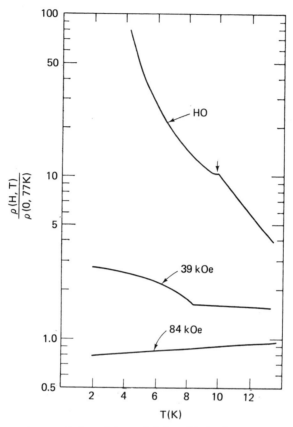

Fig. 3.5 Temperature variation of the resistivity of EuTe for several values of magnetic field (Shapira *et al.* 1972).

fields. In strong fields the moments on the Eu atoms are aligned ferromagnetically, so no spin polarons form. In the absence of a field the resistivity is greatly increased. Measurements of the Hall effect show that this is due to a change in the mobility, not in the number of carriers. The increased mass of the spin polaron affects the mobility because Anderson localization may take place in the random field due to the donors.

Somewhat similar evidence is obtained by von Molnar and Holtzberg (1973) on crystalline $Gd_{3-x}v_xS_4$, where v stands for a vacancy. The situation for $Ce_{3-x}v_xS_4$, which has the same structure, was discussed by Cutler and Mott (1968); see also Chapter 1, Section 11. An Anderson transition from hopping to metallic behaviour was observed as x (and hence the number of electrons) is changed. In the Gd compound each electron can form a spin polaron, which is heavier than a free electron; thus localization can occur more easily in the absence of a magnetic field. Figure 3.6 shows the resistivity without and with a magnetic

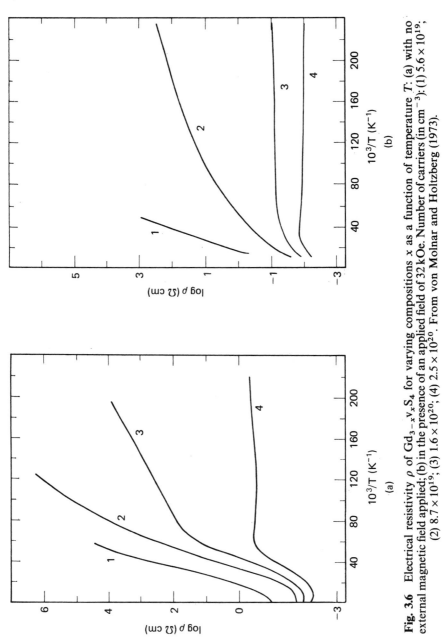

Fig. 3.6 Electrical resistivity ρ of $Gd_{3-x}v_xS_4$ for varying compositions x as a function of temperature T: (a) with no external magnetic field applied; (b) in the presence of an applied field of 32 kOe. Number of carriers (in cm^{-3}): (1) 5.6×10^{19}; (2) 8.7×10^{19}; (3) 1.6×10^{20}; (4) 2.5×10^{20}. From von Molnar and Holtzberg (1973).

field. It will be seen that the field lowers the concentration at which the Anderson transition occurs, and we ascribe this to the enhanced mass of the spin polaron.

A more recent review of the properties of this material has been given by von Molnar and Penney (1985; see also von Molnar et al. 1983, 1985). Results discussed in this article, involving the effects of disorder and electron–electron interaction, are described in Chapters 5 and 9. Briefly, the semiconductor-to-metal transition in an increasing magnetic field leads to a conductivity, at 300 mK, that increases linearly with H (von Molnar et al. 1983). This is shown in Fig. 3.7. Hopping conduction is observed with an index $v \approx \frac{1}{2}$, indicating the influence of a Coulomb gap (Washburn et al. 1984), and near the transition a temperature dependence of σ as $a + mT$, with m positive (von Molnar et al. 1985).

We shall consider in Chapter 9 the attractive interaction between two spin polarons, and the possible relationship to high-temperature superconductivity.

Very direct evidence for the existence of bound spin polarons is provided by the work of Torrance et al. (1972) on the metal–insulator transition in Eu-rich EuO. At low temperatures, when the moments on the Eu ions are ferromagnetically aligned, the electrons in the oxygen vacancies cannot form spin polarons and are present in sufficient concentration to give metallic conduction. Above the Curie temperature the conductivity drops by a factor of order 10^8, because the electrons now polarize the surrounding moments, forming spin polarons with higher effective mass.

There is also evidence for the existence of spin polarons in dilute magnetic semiconductors such as $Cd_{1-x}Mn_xTe$ with x in the range 0–0.8. This is described in the review by Furdyna and Kossut (1988).

6   A degenerate electron gas in the presence of a magnetic impurity; the RKKY interaction

Here we have in mind such materials as EuS with a comparatively high concentration of Gd atoms to give a degenerate electron gas, and a large number of "metallic" transitional-metal compounds where ions of mixed valence exist (in the latter there may be uncertainty about whether the electrons are in a conduction (4s) band or the upper Hubbard band described in Chapter 4). In such a case a new interaction term arises between the moments which is via the conduction electrons. This is the so-called RKKY (Ruderman–Kittel–Kasuya–Yosida) interaction, which is an oscillating function of distance (Ruderman and Kittel 1954, Kasuya 1956, Yosida 1957; for a detailed description see Elliott 1965). This derives from the formulae of Chapter 1, Section 5. Consider an atom with magnetic moment in a given direction; then the wave functions of conduction electrons with spin up and with spin down will vary with distance in different ways, so that

$$|\psi_\uparrow|^2 - |\psi_\downarrow|^2 = 2J_{sf} \frac{m\Omega k^2}{2\pi^3 \hbar^2} \frac{1}{r^2} \left[\frac{\cos kr}{2kr} - \frac{\sin kr}{(2kr)^2} \right].$$

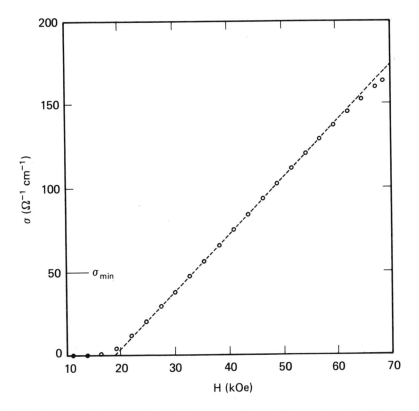

Fig. 3.7 Magnetic field dependence of the conductivity of $Gd_{3-x}V_xS_4$ at $T = 300\,mK$. An estimate of σ_{min} is also indicated. From von Molnar and Penney (1985).

The interaction between moments at a distance r will then be

$$J_{sf}^2 \frac{m\Omega^2 k^2}{2\pi^3\hbar^2} \frac{1}{r^2} \left[\frac{\cos kr}{2kr} - \frac{\sin kr}{(2kr)^2} \right]. \tag{18}$$

This coupling energy is always ferromagnetic for small values of kr, but changes sign as r increases. If the mean free path l is finite then (18) must be multiplied by $e^{-r/l}$, as shown by de Gennes (1962; see also Heeger 1969, p. 326).

For rare-earth metals the RKKY interaction leads to the well-known spiral arrangement of moments, but always with residual ferromagnetism because the average value of (18) corresponds to ferromagnetic coupling; the direct exchange (antiferromagnetic) coupling is here negligible.

For a low concentration of conduction electrons the wave vector k will be small and the RKKY mechanism will always give ferromagnetic coupling. It then becomes identical with a form of coupling first introduced by Zener (1951). Zener argued that the moments, if lined up ferromagnetically, would polarize the conduction electrons so that they would contribute a moment per atom

$$\mu \Omega J_{sf} N(E_F),$$

where $N(E_F)$ is the density of states at the Fermi energy and Ω is the atomic volume. This will give ferromagnetic coupling of energy

$$\Omega J_{sf}^2 N(E_F),$$

which for free electrons with n electron per atom is equal to

$$\Omega J_{sf}^2 m n^{1/3} / 2\pi^2 \hbar^2. \tag{19}$$

Summations of the RKKY interaction (18) over all the spins are described by Mattis (1965). Roughly, we can see that, for small k, contributions will fall off rapidly when $kr > 1$, so that from (18) the interaction energy is

$$\sim \frac{1}{6\pi^3} \frac{k^2 \Omega J_{sf}^2}{\hbar^2} \int_0^{1/k} 4\pi r \, dr \sim \frac{2}{9\pi^2} \frac{m J_{sf}^2 \Omega}{\hbar^2 k},$$

which, apart from a numerical factor, is identical with (19).

However, this model does not tell us the whole story about the properties of metals containing metallic impurities. It includes the effect of the free carriers on the moments, but not the converse effect of the moments on the carriers. It is well known that interaction with phonons leads to an enhancement of the electronic specific heat of a degenerate electron gas (Chapter 2, Section 2). That a similar effect occurs for interaction with spin waves has been suggested by several authors (Berk and Schrieffer 1966, Doniach and Engelsberg 1966), and in a detailed study Nakajima (1967) has given a treatment of the effect for ferromagnetic rare-earth metals. For interaction with phonons when the effect of phonons is large, and the number of carriers small, we have suggested (Chapter 2, Section 2) that a degenerate gas of dielectric polarons is the appropriate treatment. Under similar circumstances, particularly a low density of carriers, we now suggest that if the coupling between electrons and moments is large compared with the Fermi energy of the former then the appropriate approximation is that of a degenerate gas of spin polarons. If there are N carriers (electrons) per unit volume then the volume occupied by these polarons should be

$$\tfrac{4}{3}\pi R^3 \quad \text{or} \quad 1/N,$$

whichever is the smaller, where R is the radius calculated above, (12).

Our model is thus of a metal, with a small number of carriers in the conduction band and 3d- or 4f-moments antiferromagnetically coupled to each other by direct exchange. The carriers may be either inserted by doping or by overlap from the lower Hubbard band (cf. Chapter 4, Section 3). The most striking prediction of the model, however, is that the degenerate electron gas should have a much enhanced Pauli magnetism. Suppose that E_F is the Fermi energy of these carriers and $\zeta \, (= N/N_0)$ is the ratio of the number N of carriers to the number N_0 of metal atoms; then the susceptibility should be

$$\frac{N_0 \zeta (\mu/\zeta)^2}{E_F} = \frac{N_0 \mu^2}{\zeta E_F}, \tag{20}$$

the enhancement factor being ζ^{-1}.

The susceptibility should also increase if the coupling between spin polarons is ferromagnetic—or in other words because of the Zener coupling. The condition for ferromagnetism is that

$$\Omega N(E_F) J_{sf}^2 > J_N. \tag{21}$$

The model of a degenerate gas of spin polarons suggests that if the direct or RKKY interaction between moments is weak and E_F too great to allow ferromagnetism then the moments might all resonate between their various orientations. This would mean that it is possible in principle to have a heavily doped magnetic semiconductor or rare-earth metal in which there is no magnetic order, even at absolute zero. This possibility is discussed further in Section 8 in connection with the Kondo effect.

7 Localized moments in metals

Up to this point we have *assumed* the presence of moments on the atoms of a magnetic material, and considered their interaction with each other via exchange or the RKKY mechanism. We now consider a different problem. Suppose a metal such as copper contains a transitional-metal atom such as manganese into which the conduction electrons can penetrate, forming a Friedel virtual bound state as described in Chapter 1, Section 5. The conduction electrons in this state with opposite spins repel each other with the Coulomb repulsion e^2/r_{12}; the positive intra-atomic energy U (Chapter 2, Section 9), given by $U = \langle e^2/r_{12} \rangle$, is wholly or partly removed if the atom forms a moment. A first attack on the problem would be to determine the condition for a moment to form, which will depend on the ratio of U to the width of the Friedel virtual bound state. This was first achieved by Anderson (1961) and by Wolff (1961), and a simple adaptation of their analysis is given below. When this is done, however, we have to ask whether, at zero temperature, a metal with magnetic centres too far apart for the RKKY interaction to be appreciable does in fact have free moments. It is now known that it does not. At zero temperature the moment on the magnetic atom resonates between its various positions, owing to interaction with the conduction electrons, with a frequency ω_K known as the Kondo frequency. The ground state of the system (metal with impurity) is thus a singlet. It is only in the unrestricted Hartree–Fock approximation (i.e. that in which electrons with different spin directions can have different wave functions) that an *isolated* impurity carries a moment. However, the concept is often useful, and we shall now give the condition for a moment in this approximation.

We start with the concept of the "virtual bound state" described in Chapter 1, Section 5, where we introduced a "localized density of states" per impurity atom $N_0(E)$, which is of order Δ^{-1}, Δ being the width of the resonance. This is the addition to the overall density of states introduced by the impurity atom. We suppose first that the state is non-degenerate. Then we again introduce the Hubbard intra-atomic energy U. The quantity U has the property that two electrons in the virtual bound state (which must have antiparallel spins for non-degenerate orbitals) will contribute an amount U to the energy of the system through their interaction. We can find the moment (if any) on the atom as follows. We suppose that the number of electrons in the virtual bound state with spin up is $n_\uparrow$ and the number with spin down is $n_\downarrow$. Then each electron with its spin down has its energy raised by $n_\uparrow U$, and vice versa. The density of states for the two spin directions will be as in Fig. 3.8, the Fermi energy E_F being the same for the two spin directions. From this figure one can deduce the moment $n\mu$ on each impurity, where $n = n_\uparrow - n_\downarrow$, from the equation

$$n = \nu(E_F + \tfrac{1}{2}nU) - \nu(E_F - \tfrac{1}{2}nU), \tag{22}$$

where

$$\nu(E) = \int_0^E N_0(E)\,dE,$$

N_0 being defined for a given spin direction. Expanding (22), we find, for small n,

$$n = nUN_0(E_F) + \tfrac{1}{24}(nU)^3 N_0''(E_F)\dots,$$

so that

$$n = \left[\frac{UN_0(E_F) - 1}{\tfrac{1}{24}|N_0''(E_F)|U^3} \right]^{1/2}, \tag{23}$$

assuming N_0'' to be negative. In this case a moment will form if

$$UN_0(E_F) - 1 > 0. \tag{24}$$

Also, if variation of some parameter such as the concentration c in an alloy causes the left-hand side of (24) to vary continuously, changing sign for $c = c_0$, then the moment $n\mu$ will vary as $(c - c_0)^{1/2}$, as in Fig. 3.9.

The behaviour of the moment illustrated in Fig. 3.9 will be deduced in Section 9 also for ferromagnetic alloys such as Ni–Al. In both antiferromagnetic and ferromagnetic materials, however, a discontinuous change from a finite to a zero value is sometimes observed; examples are NiS as a function of temperature or pressure, in which a discontinuous disappearance of antiferromagnetism occurs though the electrical properties are metallic on both sides of the transition, and a discontinuous disappearance of ferromagnetism in $Co(S_xSe_{1-x})_2$ with varying x. Both are discussed in Chapter 6. Such behaviour implies that the energy E as a

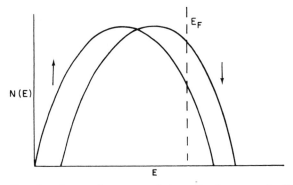

Fig. 3.8   Density of states in a magnetic impurity for two spin directions.

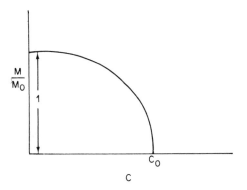

Fig. 3.9   Variation of a magnetic moment as $(c_0 - c)^{1/2}$.

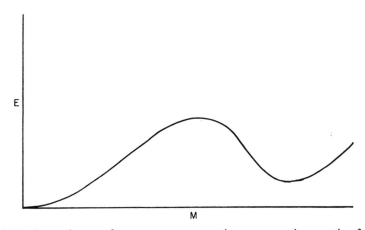

Fig. 3.10   Dependence of energy on magnetic moment in certain ferro- or antiferromagnetic materials.

function of moment is as in Fig. 3.10. We think that two minima† can occur if U depends on the degree of hybridization with the anion wave function, as we shall see in Chapter 6.

If there is a discontinuous change in the moment with varying composition or volume then there must be a "kink" in the plot of energy against volume or composition as illustrated in Fig. 4.2, so that if x is varied a two-phase region is expected. This is discussed further in Chapter 4.

In the case of a degenerate level, the formation of a moment gains the energy U and also Hund's-rule coupling between the z electrons that form the moment. The resultant energy was discussed by Klein and Heeger (1966).

We now ask how these moments can be observed. We have stated that at temperatures below the Kondo temperature there is no true moment; but at high temperatures we expect a Curie susceptibility of the form

$$\chi = \frac{N(n\mu)^2}{k_B T},$$ (25)

where n is given by (23) and N is the number of centres.

There is a very large literature on the appearance of local moments due to transitional-metal atoms dissolved in both solid and liquid metallic alloys, showing that moments occur for some concentrations of the host metal and not for others (for a summary see Heeger 1969). Some of these can be used to give evidence on whether the moment varies as $(c - c_0)^{1/2}$. The earliest results are those of Clogston et al. (1962) for the moment M on an iron atom when 1% of iron is added to transitional-metal alloys of the composition shown in Fig. 3.11. The moment is deduced from the Curie–Weiss susceptibility of type (25). The variation of M, for small values approximately as $(c - c_0)^{1/2}$, is clearly shown. Gruber and Gardner (1971) traced the appearance of moments on Fe and Mn in liquid Cu–Al as the copper concentration increases (see also Parmenter 1971). Collings (1971) showed the same for liquid Al–Sn.

We note that when moments do not appear an enhancement of the Pauli paramagnetism χ is expected of the form

$$\chi = \frac{N_0(E_F)\mu^2}{1 - U N_0(E_F)}$$

for low temperatures, going over to the Curie form when $T > T_K$, where

$$k_B T_K = \frac{1 - U N_0(E_F)}{N_0(E_F)}.$$ (26)

There are many examples in the literature of this kind of enhancement; it can be important in liquid alloys such as Ge–Co (Dreirach et al. 1972, Mott 1972c).

---

† For the ferromagnetic case a curve of this kind was proposed by Wohlfarth and Rhodes (1962), Herring (1966, p. 289) and Shimizu (1964, 1965). A discussion was given for the antiferromagnetic case by Penn and Cohen (1967).

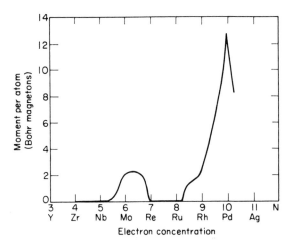

Fig. 3.11 Magnetic moment in certain alloys (Clogston *et al.* 1962).

When the theory gives a non-integral value of the moment, it is not entirely clear what value of the Curie constant should be taken in (25). Our belief is that the moment will be changing with a frequency $k_B T_K/\hbar$ from one integral number to another, and only if $T \gg T_K$ can a formula of type (25) be used. The susceptibility under these conditions will always be of the form

$$N\sum v_s (n_s \mu)^2 / k_B T,$$

where v_s is the weight in the wave function of the state with n_s d-electrons, n_s being integral.

The possibility that in an alloy different atoms have different moments (or occupation numbers) was stressed by Jaccarino and Walker (1965), who considered that moment formation in a uniform host should be a discontinuous process, so that moments are always integral, and that in alloys, as shown by the results of Clogston *et al.*, an apparent continuous change is due to the different surroundings of each transitional-metal atom. Brog and Jones (1970) gave NMR evidence in favour of this. We do not think though that this is necessarily so in all cases. A non-integral moment, however, can mean two things:

(a) a rapid variation with time in the occupation number (as in **CuNi**—thus, if $x\mu$ is the moment, x is the number of atoms having moment μ at a given instant of time, and so we expect at high T

$$\chi = \frac{Nx\mu^2}{k_B T};$$

(b) a situation such as that described by Grüner and Mott (1974) in which there is a slow Kondo fluctuation between two g-values (cf. Section 8).

8   The Kondo effect

We have seen that a magnetic impurity may or may not produce a magnetic moment in a metal. If it does, and if the moment is that of a single spin ($s=\frac{1}{2}$), then at zero temperature an effect will occur in which a conduction electron from near the Fermi energy jumps onto the impurity, with spin antiparallel to that of the electron already there, this happening with a low frequency ω_K, and then the other electron jumps (rapidly) back to a state at the Fermi energy of the metal. The ground state of the system is thus a singlet,† and the magnetic susceptibility resulting from a dilute concentration of impurities as a function of T is as in Fig. 3.12, the linear increase of χ^{-1} setting in for $k_B T \gtrsim \hbar \omega_K$. Effects of this phenomenon are an enhancement of the term γ in the electronic specific heat γT and in the Pauli susceptibility, and a drop in the electrical resistivity ρ as T increases, leading to a minimum in ρ, before scattering by phonons becomes the major term affecting it. This was the phenomenon originally explained by Kondo (1962), and these effects are by no means limited to impurities for which $s=\frac{1}{2}$.

In addition, there are certain metals involving 4f- or 5f-states, formerly called "Kondo metals" and now described as heavy-fermion materials, which have a very large electronic specific heat and may be described as crystals in which every rare-earth metal is envisaged, coherently, in a Kondo-type spin flip.

Many papers have been published on the theory of the Kondo effect, including some exact solutions. We recommend the 260 page review by Tsvelich and Weigmann (1983). Our aim in giving a simple non-mathematical account is to point out the similarity between the enhancement of the effective mass that occurs in crystalline metallic systems near to the conditions for a Mott transition (Chapter 4), and also to address the possible effects of free spins in doped semiconductors near the transition (Chapter 5).

Our first point is that, in a material containing per unit volume N magnetic impurities with $s=\frac{1}{2}$, the proportion η that are doubly occupied is given by

$$\eta = \exp(-E_K/\tfrac{1}{2}\varDelta), \tag{27}$$

where $\varDelta$ is the width of the Friedel virtual bound state. E_K is the energy required to take an electron from the Fermi surface of the metal onto the magnetic impurity; alternatively, it can be the energy required to take the electron from the impurity to the Fermi surface, because a spin flip can equally well be initiated that way—for E_K in (27) we should take whichever is the smaller. The two types of spin flip are illustrated in Fig. 3.13.

The frequency of the spin flip is

$$\hbar \omega_K = \varDelta \exp(-E_K/\tfrac{1}{2}\varDelta). \tag{28}$$

We now assert that the range of energies that give a peak in $|\psi|^2$, originally of width $\varDelta$, is narrowed by a factor η. Thus the contribution of each magnetic

---

† This is not so for $s>\frac{1}{2}$, the ground state having weight $2s$ (Mattis 1967).

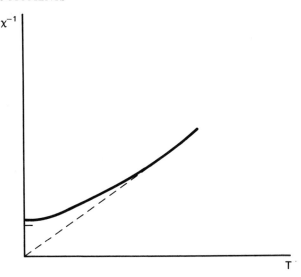

Fig. 3.12 Dependence of the reciprocal of the magnetic susceptibility χ^{-1} on temperature for a material containing Kondo centres.

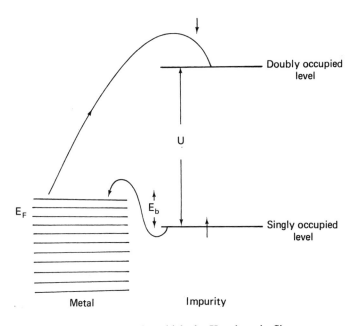

Fig. 3.13 The two processes by which the Kondo spin flip can occur.

impurity to the paramagnetism is $\mu^2/\hbar\omega_K$ below $\hbar\omega_K/k_B$ and $\mu^2/k_B T$ above. In the same way, the electronic density of states for N impurities per unit volume is enhanced. When this is greater than the value without impurities, the specific heat is γT, where $\gamma = (N/\hbar\omega_K)k_B$.

We can now see why the conductivity drops with increasing temperature. The conductivity can be written (see Chapter 1, equation (31))

$$\sigma = \int \sigma(E) \frac{\kappa}{\partial E} \, dE, \tag{29}$$

and since, as described above, the resonance width is narrowed by the factor ξ, giving $T_K = \xi \Delta/k_B$, we expect that owing to the width of the function $\partial f/\partial E$ (Mott and Jones 1936, Chap. 7),

$$\rho = \rho_0 \left[1 - \frac{\pi^2}{6} \left(\frac{T}{T_K} \right)^2 \cdots \right]. \tag{30}$$

Caplin and Rizzuto (1968) used this argument to describe their observation that in dilute alloys of Mn and Cr in Al:Cr the resistance has the form

$$\text{constant} \times \left[1 - \left(\frac{T}{\Theta} \right)^2 \cdots \right],$$

with $\Theta = 530\,\text{K}$ for Mn and $1200\,\text{K}$ for Cr.

When $T > T_K$ Kondo's derivation of the logarithmic decrease in resistivity started with the Hamiltonian (cf. (11))

$$-JSs(r), \tag{31}$$

where S is the moment on the ion, $s(r)$ the distributed moment of the conduction electron and J a coupling constant. His derivation, which will not be described here, gave a logarithmic term in the resistivity

$$\rho_0 \left[1 - JN(E) \ln \left(\frac{k_B T}{\Delta} \right) \right], \quad T > T_K. \tag{32}$$

The approximation breaks down at the Kondo temperature T_K, when the term in square brackets vanishes, so that this approach gives for T_K

$$k_B T_K \approx \Delta \exp \left[-\frac{1}{JN(E)} \right]. \tag{33}$$

The resistance in the transitional region is discussed by Rivier and Zlatić (1972). Figure 3.14 shows how, according to these authors, spin flip and elastic scattering contribute to the resistivity.

The quantity J that should be used in (31) to calculate, for instance, exchange scattering is, according to Schrieffer and Wolff (1966),

$$2I_{sd}^2 \left(\frac{1}{E_a} + \frac{1}{E_b} \right). \tag{34}$$

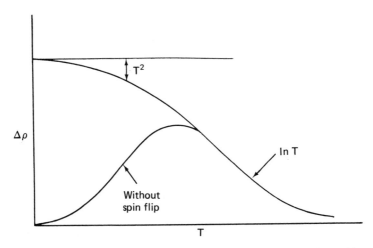

Fig. 3.14 Excess resistivity with and without spin flip due to transitional-metal impurity showing Kondo effect (schematic).

This can be compared with the formula (7) for antiferromagnetic coupling. The factor 2 arises because the difference between the spin-up and spin-down energies in (31) is $\frac{1}{2}J$. These authors inserted (34) in (33) to obtain J_{K}. The argument given above, however, suggests that one should insert in (33) either

$$\frac{I_{\mathrm{sd}}^2}{E_a} \quad \text{or} \quad \frac{I_{\mathrm{sd}}^2}{E_b},$$

whichever is the larger. If this is done, making use of the formula (Chapter 1, equation (26))

$$\varDelta = \pi I_{\mathrm{sd}}^2 \varOmega N(E),$$

then (32) gives

$$k_{\mathrm{B}}T_{\mathrm{K}} = \varDelta\, e^{-\pi E_{\mathrm{K}}/\varDelta}, \tag{35}$$

which is as close to (28) as our approximation would lead us to expect. (The term J will produce the normal exchange (Korringa) spin-flip with frequency $(J_1/E_{\mathrm{F}})^2 k_{\mathrm{B}}T/\hbar$; this spin-flip period will lead to a temperature-dependent paramagnetic Curie temperature in the susceptibility, so we may set

$$\chi = \frac{N\mu^2}{k_{\mathrm{B}}T[1+(J/E_{\mathrm{F}})^2]+\varTheta},$$

which may account for the observed μ-values, which are sometimes rather less than those expected.)

Figure 3.15 (Heeger 1969, p. 306) shows the added resistivity due to iron-group impurities in gold. The low-temperature values, for which scattering cross-sections of order a^2 occur (the "unitarity limit"), include Kondo scattering. At room temperature, $k_{\mathrm{B}}T$ is too great for most of the electrons near E_{F} to resonate

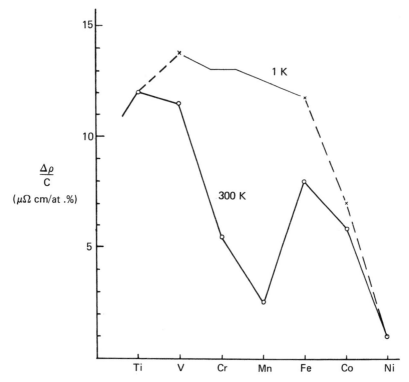

Fig. 3.15 Added resistivity of iron-group impurities in gold at 1 K and 300 K (Heeger 1969).

with the Kondo state, and the one-electron Friedel treatment is valid. The minimum at Mn occurs because η_2, by the Friedel sum rule, is equal to π, so that $e^{2i\eta_2} - 1$ vanishes and there is no d-scattering.

Two other points about Kondo scattering will be important for our subsequent discussion. For $T < T_K$ a large negative magneto-resistance is expected; if $H\mu > k_B T_K$ then the moment is now fully orientated and only conduction electrons of one spin direction can resonate with it. This effect has been observed in CuMn by Monod (1967). The second is the extensive work showing that pairs of atoms on adjacent sites, or groups of more than two, usually give a lower Kondo temperature that single atoms (see e.g. Saint-Paul et al. (1972) and Tholence and Tournier (1970); for a general theoretical discussion see Matho and Béal-Monod (1972, 1973)). The lowering of the Kondo temperature depends on ferromagnetic coupling between the atoms.

The question of the existence of the Kondo effect in amorphous systems is of interest for the considerations of Chapter 5. There is no theoretical reason to suppose that the Kondo temperature will be greatly affected; on the other hand, the short mean free path l should cut down the RKKY interaction, which, for distances r greater than l should fall off as $e^{-r/l}$ (de Gennes 1962). In alloys

Souletie and Tournier (1969) have not been able to observe much change, but Liang and Tsuei (1973) have observed a Kondo resistivity minimum in an amorphous $Ni_{41}Pd_{41}B_{18}$ alloy containing Cr; as expected, they find that the d–d spin correlation between magnetic atoms is weaker in amorphous alloys than in the corresponding crystals.

9 Heavy-fermion materials

We now discuss briefly the materials known formerly as "Kondo metals", and now usually called "heavy-fermion materials" (for a review of their properties see Coles 1987, Fulda 1988, Fulda et al. 1988). Essentially these are compounds or metallic oxides in which conduction is partly in a 4f-band of a rare earth or the 5f-band of metals such as uranium. These include $CeAl_3$, UAl_2 and $CeAl_2$. Some become magnetically ordered or superconducting at low T. Such materials can have various properties. The rare-earth metals are normally magnetic, with an integral number of electrons in the 4f-states, giving moments in some kind of canted antiferromagnetic array. The mixed-valence materials such as SmS (Chapter 1, Section 16) have a non-integral number of electrons in an f-band, the Fermi surface being partly d- or sp- and partly 4f-like; they also have a large electronic specific heat. The heavy-fermion materials, on the other hand, are nearer in character to the rare-earth metals, but each 4f-atom acts coherently with its neighbours as a Kondo centre; one has to say that the Kondo interaction is stronger than the RKKY coupling between the moments that produces the antiferromagnetism in the rare earths or a spin glass in the materials considered in Section 8. The enhanced paramagnetism and specific heat correspond, and are the same, in the case of $CeAl_3$, as the enhancement produced in Al by a small concentration of cerium atoms, multiplied by the actual number. Figure 3.16 shows its resistivity together with that of $CeAl_2$; other alloys of cerium and aluminium were measured by Buschow and van Daal (1970). The Ce ion is in the state $J = \frac{5}{2}$ and shows Curie–Weiss behaviour ($\mu_{eff} = 2.53\mu_B$, $\Theta = -40 K$) down to $10 K$. It will be seen that the other alloy shows a sharp resistivity peak (a Curie or Néel temperature), with a Kondo drop at higher T. For $CeAl_3$, on the other hand, ρ increases as T^2 and then drops when $k_B T$ becomes greater than the narrowed f-band width.

10 Transitional metals and their alloys

In the magnetic transitional-metal compounds to be described in subsequent chapters the conduction electrons lie wholly in a d-band. This does not of course mean that there is no hybridization with 4s-electrons; such hybridization must always be present to some extent, and will be responsible for any observable Knight shift. It does mean, however, that the d-band remains separate from the conduction band and contains an integral number of electrons per atom. As we

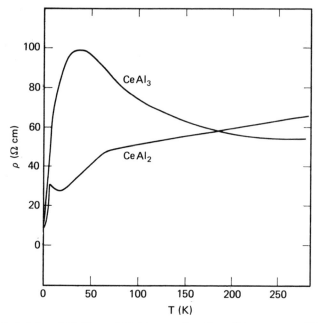

Fig. 3.16   Resistivity of $CeAl_3$ and $CeAl_2$ as a function of temperature (Buschow and van Daal 1970).

shall see again in Chapter 4, a band containing an integral number of electrons must exceed a certain width if it is to be metallic. In transitional-metals, on the other hand, a d-band and an s-band overlap, as discussed in Chapter 1, Section 2.2.2 and illustrated in Fig. 1.7. Two methods of calculating d-bands were used in early work. One is the tight-binding method for the d-band, followed by hybridization with a 4s-band, and the other makes use of the phase shifts of the 4s-electrons (Heine 1969). Their relative advantages are discussed by Friedel (1973). In any given case the d-band width may be due primarily to d–d overlap or to s–d hybridization. (For recent reviews see Edwards (1979), Pettifor (1986) and Cottrell (1988).)

For the metals Co, Ni and Pd and perhaps others it appears to be a good approximation to assume, in spite of the hybridization, that part of the Fermi surface is s-like with $m_{eff} \sim m_e$, and part d-like with $m_{eff} \gg m_e$. The current is then carried by the former, and the resistance is due to phonon-induced s–d transitions. This model was first put forward by Mott (1935) and developed by many other authors (e.g. Coles and Taylor 1962); for reviews see Mott (1964) and Dugdale and Guénault (1966). Applications of the model have also been made to ordered alloys of the type Al_6Mn, Al_7Cr by Grüner et al. (1974), where the width Δ of the d-band is the same as it would be for an isolated transitional-metal atom in the matrix, but most of the Fermi surface is assumed to be (s–p)-like. The behaviour of the disordered Pd–Ag alloy series is particularly interesting. The 4d-bands of the two constituents are well separated, as shown particularly by the photoemission work of Norris and Myers (1971). As Ag is added to Pd, the holes in the 4d-band fill up, and from susceptibility, resistivity and thermopower measurements the band appears to be full at about 50% Ag. We then envisage a d-band in which, owing to disorder, $\Delta k/k \sim 1$ as described in Chapter 1, Section 6, though this does not mean that the Fermi *energy* is blunted. The photoemission studies confirm that $\Delta k/k$ is small for the d-band, since peaks depending on selection rules disappear. There does not appear to be any marked tail to the upper limit of the d-band.

At small concentrations of Pd there will be a conduction-band resonance at each transitional-metal atom, and the resistance depends on the phase shifts. For Ag–Pd the resonance only just extends over the Fermi energy, giving very small residual resistance and specific-heat enhancement; for Cu–Ni they are much larger.

The intermediate range of concentrations between those at which resonant scattering and s–d transitions are appropriate has not been fully explored, except in the CPA approximation (Stocks *et al.* 1973), which does not give the mean free path. For liquid transitional metals the present author (Mott 1972d) has suggested that one must introduce two mean paths, l_s for the s-electrons and l_d for the d-electrons, that $l \sim a$ for the latter (as in the alloy) and that s–d transitions are appropriate to describe the resistance. Other authors have described the resistance in terms of a single mean free path, determined by the resonant scattering of the s-electrons by the d-shells (Evans *et al.* 1971).

11 Ferromagnetic and nearly ferromagnetic metals

This section does not attempt a survey of this large subject. Its aim is only to show briefly how metallic Ni, Co and Fe differ from the transitional-metal oxides that are of particular interest to us in connection with the metal–insulator transition. The former are in our view in a sense more complicated, because the number of electrons in the d-band is in all cases non-integral, while in metallic oxides (such as V_2O_3 under pressure and CrO_2) this is not the case.

Discussions of ferromagnetism in these metals can start from two models. In that first put forward by Friedel *et al.* (1961) one first asks whether individual atoms carry moments using a criterion of type (24), and, if so, whether the interaction between them—in principle of RKKY type—is ferromagnetic or antiferromagnetic. This method was developed further by Alexander and Anderson (1964) and Moriya (1965). The other method goes back much further and has its origin in the work of Slater (1936), Stoner (1947) and Wohlfarth (1949). Here we begin with a calculation of the density of states $N(E)$ and then introduce an energy U_{eff}, which is the difference between the energy of a pair of electrons in Bloch states when the spins are parallel and that when they are antiparallel. U_{eff} is not the same as the Hubbard U, as we shall see below. This gives for the condition for ferromagnetism (Friedel *et al.* 1964), exactly as in (24),

$$U_{eff}\Omega N(E_F) > 1. \tag{36}$$

Formula (36) represents the condition that the energy will be lowered by a *small* polarization of the spins of electrons at the Fermi surface. The condition for a large polarization will involve the value of $N(E)$ over a considerable range of E. This formula was generalized for this case by Shimizu and Katsuki (1964) and Shimizu (1964, 1965) (cf. also Mott 1964, p. 346, footnote).

As regards the magnitude of the moment on each atom, the analysis is similar to that of Section 7 for the moment on a single atom for a virtual bound state. However, we cannot now assume that the Fermi energy is not shifted when our equations are expanded to third order in U. Unless $N'(E_F) = 0$, it is, and the moment becomes

$$M = M_0 \left[\frac{\Omega U_{\text{eff}} N(E) - 1}{-\frac{1}{24}N'' + N'^2/8N} \right]^{1/2}, \tag{37}$$

provided that the numerator is positive (see e.g. Wohlfarth and Rhodes 1962). The same square-root dependence of M as that illustrated in Fig. 3.9 is predicted. Figure 3.17 (due to Wohlfarth 1971) shows results described by that author for the system Ni–Al. Both M^2 and T_c^2 vary as $c - c_0$. Similar results are obtained for Ni–Pt (Besnus and Herr 1972). This behaviour, however, is *not* shown by Cu–Ni alloys, where the moment drops *linearly* with increasing copper content. Mott (1935) and Mott and Jones (1936) interpreted this as being due to a filling up of the 3d-band of nickel by the extra electron from the copper atom, until an equilibrium value of ~ 0.5 4s-electrons is reached. This interpretation works well for the paramagnetic alloy system Ag–Pd. Coles (1952) pointed out, however, that the strong paramagnetism of **Cu**Ni alloys makes this interpretation impossible for this system. For all concentrations of Ni, d-holes remain. The linear behaviour is unexplained, but may be connected with the spin fluctuations observed over long distances in the alloy (see Lederer and Mills 1968). For ferromagnetic alloys there is much evidence that spin polarization clouds form in local nickel-rich regions. Kouvel and Comly (1970) find that paramagnetic clusters coupled antiferromagnetically form on the paramagnetic side. For a theoretical discussion see Fibich and Ron (1970).

The term U_{eff} can either saturate the moments (as in nickel), or give a smaller moment according to (36). The material to which the second model is usually applied is the weak ferromagnetic $ZrZn_2$ (Knapp *et al.* 1971), where there appears to be a peak in the density of states at the Fermi energy, giving N'' strongly negative so that the denominator in (37) is positive.

If U_{eff} is not great enough to produce ferromagnetism then, as for isolated impurities, the Pauli susceptibility is increased by the Stoner enhancement factor K_0^{-2}, where $K_0^2 = 1 - N(E)\Omega U_{\text{eff}}$. This factor can become indefinitely large. There are striking examples of this behaviour. One is provided by the alloy systems Ni–Al and Ni–Ga, which lose their ferromagnetism at the compositions Ni_3Al or Ni_3Ga, as illustrated in Fig. 3.17. The susceptibilities of Ni–Ga near the transition obtained by de Boer *et al.* (1969) are shown in Fig. 3.18. It will be seen that, as the enhancement factor increases, the $1/\chi$–T curves are shifted downwards without

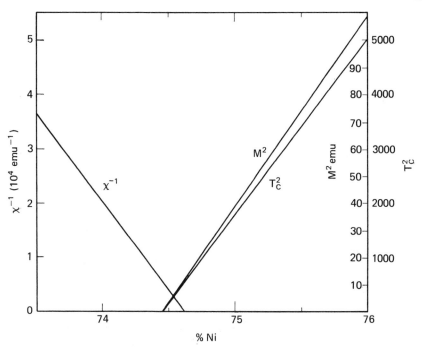

Fig. 3.17 Squares of moment and of T_c and reciprocal of susceptibility plotted against composition in Ni–Al alloys. Reproduced from results of Wohlfarth (1971).

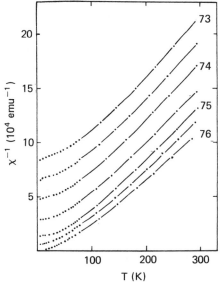

Fig. 3.18 Reciprocal of paramagnetic mass susceptibility χ^{-1} of $Ni_{100-x}Ga_x$ for various compositions near Ni_3Ga as a function of T (de Boer *et al.* 1969). Note the behaviour, here ascribed to paramagnons, at compositions 73 and 73.5 at the lowest temperature.

change of form. More detailed confirmation of this behaviour was given by Schinkel et al. (1973). This is what one would expect from the simple application of a Stoner enhancement factor to single-particle excitations, so that (Mott and Jones 1936, p. 186)

$$\chi = 2\mu^2 K_0^{-2}[N(E) + \tfrac{1}{6}\pi^2 k_B^2 T^2 N''(E) + \ldots]. \tag{38}$$

This is the conclusion obtained by the most recent development of the Stoner–Wohlfarth theory (Edwards and Wohlfarth 1968).

On the other hand, a series of papers (Berk and Schrieffer 1966, Doniach and Engelsberg 1966, Mills and Lederer 1966) predict that, just as magnons are possible excitations of a metallic ferromagnet, so "paramagnons" or spin fluctuations are well-defined excitations of a material where K_0 is small so that it is nearly ferromagnetic. Renormalization of the single-particle excitations shows that the formation of paramagnons does not increase χ_0. As regards the temperature dependence, Béal-Monod et al. (1968) showed that, for the part of the paramagnetism that can be ascribed to paramagnons,

$$\frac{\chi}{\chi_{T=0}} = 1 - \frac{\pi^2}{24} 3.2 K_0^{-2} \left(\frac{T}{T_F}\right)^2, \tag{39}$$

and so χ behaves quite differently from (38). But in neither a weak ferromagnet nor a strong paramagnet can the number of possible magnons be large, because those of short wavelength are critically damped. We suppose then that their contribution dies away at $T \sim K_0^2 T_F$, and the total susceptibility curve appears as in Fig. 3.19. Some sign of this predicted paramagnon behaviour appears in Fig. 3.18.

We note that Stoner enhancement does *not* give an equivalent enhancement of either γ (the coefficient in the electronic specific heat γT) *or* the Baber scattering (Chapter 2, Section 6). Paramagnons, on the other hand, can give an enhancement of γ, but *not* one that goes to infinity as $K_0 \to 0$. Experiments by de Dood and de Chatel (1973) have found some anomalies in γ in Ni_3Al and Ni_3Ga, not in accord with those predicted; they conclude there is no paramagnon effect.

In PdNi, which becomes ferromagnetic at concentrations near 3% Ni, ρ increases by only about 50% from the value for pure Ni, while χ increases by a factor of about 4 (Chouteau et al. (1968a). For a theoretical discussion of this work see Chouteau et al. (1968b). The specific heat was also measured by Schindler and Mackliet (1968).

Perhaps the most sensitive test for the presence of paramagnons is the predicted large T^2 term in the resistivity; if this occurs without a correspondingly large increase in γ then paramagnons rather than Baber scattering are indicated. Figure 3.20 (from Tari and Coles 1971) shows the behaviour of the constant A in Pd–Ni, where

$$\Delta\rho = AT^n.$$

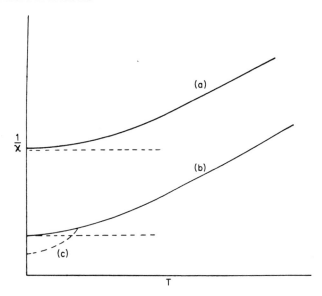

Fig. 3.19 Expected effect on susceptibility due to paramagnons (schematic): (a) without enhancement; (b) with Stoner enhancement; (c) effect of paramagnons.

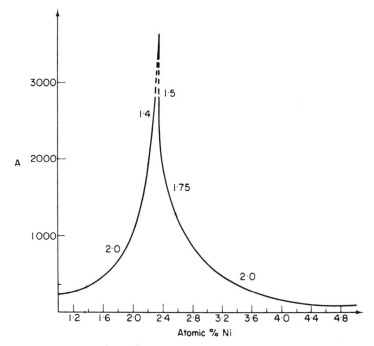

Fig. 3.20 Plot of A (in $\Delta\rho = AT^2$) against Ni concentration in Pd–Ni (Tari and Coles 1971). The index n in T^n is shown. Note that Mathon (1968) predicted $n \approx \frac{5}{3}$ near the peak.

n is not accurately equal to two, and the values are shown in the figure. For Ni_3Al and Ni_3Ge a discussion is given by Fluitman *et al.* (1973), the resistivity here being bigger than Baber scattering can account for.

To summarize, then, Stoner enhancement can increase the Pauli susceptibility without limit. The temperature dependence does not become large, except for the part due to paramagnons, when a significant rise is expected at low T; paramagnons do not produce a large value of γ, but can produce a large T^2 resistivity term (for a summary of experimental work on paramagnons in nearly ferromagnetic systems see de Chatel and Wohlfarth 1973). In Chapter 4 we discuss the behaviour of a highly correlated gas (a nearly antiferromagnetic metal) and find that here, in contrast, γ is enhanced about as much as χ.

If the material is ferromagnetic then the entropy, susceptibility and resistance at temperatures just above the Curie point are to be calculated in much the same way (i.e. in terms of spin fluctuations). A treatment of this problem starting from the Stoner–Wohlfarth model is due to Moriya and Kawabata (1973).

We now turn to the nature of the enhancement term U_{eff} in (36) and (37). We note that we are dealing with a band with a non-integral number of electrons per atom, which therefore *must* be metallic. Kanamori (1963) showed that such a band may be ferromagnetic on the Hubbard model. Here we consider first the case when the atomic orbitals are non-degenerate. The quantity U_{eff} occurring in (36) is *not* the same as the intra-atomic energy defined by (3) and used in calculating antiferromagnetic interaction. Herring (1966), following Kanamori, gave a detailed discussion of its magnitude. A useful approximation is to set

$$U_{eff} = \frac{U}{1 + \beta \Omega U N(E_F)}, \tag{40}$$

where β is a constant. (For papers determining β from the dependence on pressure of the magnetic constants see Wohlfarth and Bartel (1971), Smith *et al.* (1971), Beille *et al.* (1973) and references therein.) One can say that if U is very large then the electrons cannot get on to the same atom; the ferromagnetism then arises because, if spins are parallel, the electrons keep away from one another owing to the symmetry of the wave function; but if they are antiparallel, keeping apart costs them kinetic energy. If, on the other hand, U is small, two antiparallel electrons can come on to the same atom, but it costs them an energy U. Lonzarich and Taillefer (1985) have reviewed the subject of spin fluctuations in such materials.

To summarize, we again want to make a sharp distinction between the enhancement in nearly ferromagnetic metals like Ni_3Al, where the Stoner enhancement factor can increase χ without limit but not $d\chi/dT$ or γ, and the enhancement in nearly antiferromagnetic metals to be discussed in Chapter 4, where χ, γ and $d\chi/dT$ are all enhanced. In the former case, all models involve either a non-integral number of electrons in the band or degenerate orbitals. In the latter, two electrons on the same atom necessarily have antiparallel spins.

12 Amorphous antiferromagnets and spin glasses

12.1 Introduction

In Section 7 we pointed out that in dilute alloys of the **CuMn** type, if the RKKY interaction between the Mn atoms is greater than the Kondo energy $\hbar\omega_K$, we expect a situation in which each Mn moment is frozen in position relative to all the other moments. Such materials have been called "spin glasses". The sign of the RKKY interaction changes with distance, so it is hardly correct to call such materials amorphous antiferromagnets. However, they do not appear to be ferromagnetic, and can perhaps be included with materials in which the coupling is wholly or mainly antiferromagnetic. These include the following.

(a) Doped semiconductors in which the concentration is on the insulator side of the metal–insulator transition (see Chapter 5): here the direct exchange as described in Section 2 may be sufficient to provide the interaction, though for more distant centres the interaction may be a modification of RKKY that results from a band gap, oscillating with distance but falling off exponentially instead of as r^{-3}. Treatments that show this are given by Bloembergen and Rowland (1955) and Cullen (1968).

(b) Glasses containing transitional-metal ions, such as partially oxidized V_2O_5–P_2O_5: here the interaction will be through the oxide ions as already described.

In all of these materials there can be no long-range antiferromagnetic order. None the less, at zero temperature there will be a fixed direction of the moment on each atom. The kind of arrangement of moments that we postulate is shown in Fig. 3.21. The whole arrangement can rotate, as for a crystalline antiferromagnet (Section 2), but the times involved will be proportional to the number of atoms, and the rate of rotation will probably be negligible. There are many questions regarding such systems that have not been fully answered, particularly those that involve the excited states. We first remark that the Anderson (1958) localization theorem can be applied—in fact it was first put forward to apply to this case—and if there is a large enough disorder, i.e. if $\Delta J > 2J$, where J is the coupling between moments, then we should expect all excitations to be localized. This means that, in a typical excitation, there will be comparatively large changes of direction of the moment over a small volume. On the other hand, if $\Delta J/J < 2$ then the lowest states will be localized but the higher ones extended; this means that the higher ones will be strongly scattered spin waves, for which k is not a good quantum number.

Buyers *et al.* (1971) discussed evidence for the existence of Anderson-localized spin states in antiferromagnets in which there is disorder. They investigated the substitutionally disordered materials $K(Co,Mn)F_3$ and $(Co,Mn)F_2$, finding two branches of propagating spin waves corresponding to the two constituents and

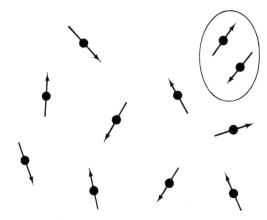

Fig. 3.21 Hypothetical arrangement of spins in an amorphous antiferromagnetic; and moments encircled form a cluster.

localized states where the branches overlap. Theoretical accounts of this situation were given by Lyo (1972) and Economou (1972). But this work does not in our view give firm evidence for localization, but only for the breakdown of the k-selection rule, or in other words $\Delta k/k \sim 1$. Evidence for localization can only come from the absence of spin diffusion.

12.2 Specific heat

Marshall (1960) was the first to point out that the linear specific heat observed in dilute solutions of transitional metals at low temperatures could be given a simple explanation. He used a theory analogous to Einstein's specific-heat model in which each moment is acted on by molecular field H_0. For simplicity, taking the case where for the angular momentum $s = \frac{1}{2}$ with a moment equal to one Bohr magneton, and since the field H_0 (being of RKKY type) can change sign, the energy of the moment at temperature T is

$$\int_{-\infty}^{\infty} H_0\mu \left[\tanh\left(\frac{\mu H_0}{k_B T}\right) - 1 \right] N(H_0)\,dH_0, \tag{41}$$

where $N(H_0)$ is the distribution function for such fields; and if N is constant over a range $k_B T/\mu$ *around a zero value of* H_0 then this is proportional to T^2, giving a linear specific heat. The theory was discussed in some detail by Brout (1965). It was criticized by Anderson *et al.* (1972) on the grounds that it is a sort of Ising model, in which one considers only s_z and not s_x and s_y. These authors gave an alternative description in terms of their discussion of the linear specific heat due to the phonon spectrum of glasses.

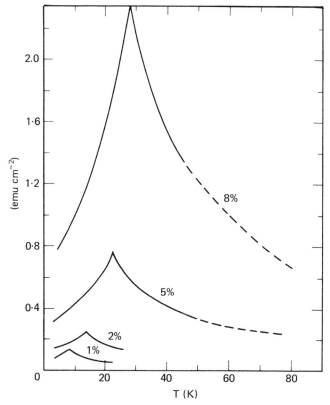

Fig. 3.22   Susceptibility of **AuFe** alloys as a function of temperature (Cannella and Mydosh 1972).

12.3   The Néel temperature

Our next problem is to ask whether a Néel temperature can exist for an amorphous antiferromagnet. Obviously a Néel temperature cannot imply a sudden disappearance of the long-range order, but we can define a Néel temperature by the value of $\langle s_z \rangle$, $\langle \ \rangle$ denoting a time average over a time large compared with spin-wave frequencies but small compared with the very long times for rotation of the whole assembly of moments. A Néel temperature is defined as the temperature at which $\langle s_z \rangle$ vanishes for all moments. It seems to us likely that if $\langle s_z \rangle$ is non-zero for some moments than it will be non-zero for all, and the question is whether it is zero at all finite temperatures, or only above some definite value T_N. Our conjecture is that if there is anisotropy at each moment, so that only a finite number of spin directions are allowed, then a finite T_N is expected. In doped semiconductors where the orbits are large it may be very low indeed.

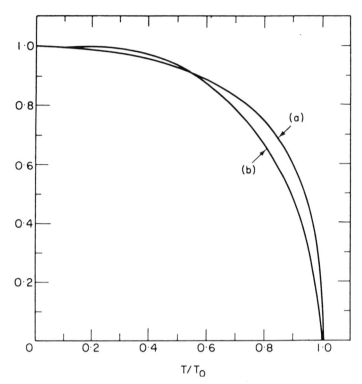

Fig. 3.23   Ground-state splitting of Fe in Au determined by Mössbauer effect (Violet and
Borg 1966): (a) observed; (b) Brillouin function for $s = 1$.

In a series of papers on cobalt and other phosphate glasses, Simpson (1970)
and Simpson and Lucas (1971) showed that there is no sign of a Néel temperature
down to ~ 1 K while the corresponding crystals show a Néel temperature near
20 K. In fact the $1/\chi$–T curve shows increased slope at low temperatures. Simpson
pointed out that the theorem of Ziman (1952) (see Section 2)—that nearest-
neighbour interaction cannot give antiferromagnetism for spherical orbitals—
may be applicable here; the orbitals are not spherical but are oriented at random.

In spin glasses, on the other hand, there is evidence of sharp Néel
temperatures. An example is the system **Au**Fe investigated by Cannella *et al.*
(1971) and Cannella and Mydosh (1972); their measurements of the susceptibility
are shown in Fig. 3.22. Mössbauer investigations on Au : Fe give similar results
(Violet and Borg 1966), the local field tending to zero at a sharp temperature T_0.
Figure 3.23 shows that the ground state splits as a function of temperature, T_0
increasing with iron concentration. Another example (VO_x) is discussed in
Chapter 6.

A sharp transition is not always seen. Figure 3.24 shows the results of Loram
et al. (1971) on the resistivity of **Au**Mn for several concentrations of manganese,
including 0.01%, where the RKKY coupling energy lies below $k_B T$, and 0.5%, for

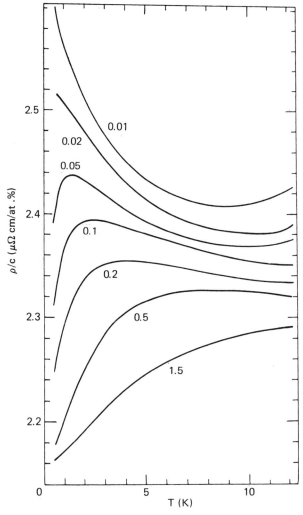

Fig. 3.24 Excess resistivity ρ/c for Mn in gold. The numbers next to the curves show the concentration of Mn in atomic % (Loram *et al.* 1971).

which it lies above it. It is perhaps relevant that the Mn $3d^5$ ion is spherically symmetrical.

There is no adequate theory of the Néel temperature of a random distribution of centres in a dilute alloy, of indeed one exists. For higher concentrations of the magnetic matrix, with the assumption that only nearest neighbours interact, there is considerable theoretical work, giving a "percolation limit", the concentration c_0 at which long-range order disappears. The behaviour of T_N is as $(c - c_0)^{1.2}$. For details see Brout (1965), Elliott and Heap (1962) and Klein and Brout (1963).

In an amorphous ferromagnet (a sputtered film with the composition $Tb_{0.33}Fe_{0.67}$) Rhyne *et al.* (1972) used neutron diffraction to demonstrate a random direction of moments with average ferromagnetic orientation. This material has a fairly sharp Curie temperature in the range 380–390 K, the moment below this temperature reaching saturation at about 50 kOe.

12.4   The susceptibility–temperature curve

There are several treatments of molecular-field type dealing with the susceptibility–temperature curve of amorphous antiferromagnets, which do not deal specifically with the problem of the Néel temperature. These include those of Hasegawa and Tsuei (1970) with applications to amorphous Mn–P–C, Simpson (1970) and Simpson and Lucas (1971) for iron and vanadium phosphate glasses and Quirt and Marko (1971) dealing with the cluster approach in doped semiconductors. In such situations some moments will be weakly bound to their neighbours and should give a large susceptibility at low T. In spin glasses where the short-range interaction is ferromagnetic the same will be true of clusters. In either case, as the temperature is raised, more and more moments are free to rotate, and the $1/\chi$–T curve will thus be concave to the T-axis. Such behaviour is widely observed.

It is worth pointing out here, however, that $1/\chi$–T curves of this form are also predicted for a highly correlated electron gas (Chapter 4, Section 9).

4

Metal–Insulator Transitions in Crystals

1 Introduction

The present author (Mott 1949, 1956, 1961) first proposed that a crystalline array of one-electron atoms at the absolute zero of temperature should show a sharp transition from metallic to non-metallic behaviour as the distance between the atoms was varied. The method used, described in the Introduction, is now only of historical interest. Nearer to present ideas was the prediction (Knox 1963) that when a conduction and valence band in a semiconductor are caused to overlap by a change in composition or specific volume, a discontinuous change in the number of current carriers is to be expected; a very small number of *free* electrons and holes is not possible, because they would form excitons.

The next step in our understanding of the transition for one-electron centres was the work of Hubbard (1964a, b), who introduced the Hamiltonian

$$H = \sum_{i,j} t_{ij} a_{i\sigma} a_{j\sigma} + U \sum_i n_{i\uparrow} n_{i\downarrow}. \qquad (1)$$

Here U is the intra-atomic interaction $\langle e^2/r_{12} \rangle$ defined in Chapter 4 and t, the hopping integral, is equal to $B/2z$, where B is the bandwidth and z is the coordination number. The suffixes i and j refer to the nearest-neighbour sites, and $a_{i\sigma}$ is the creation operator for site i. The suffix σ refers to the spin direction. Hubbard found that a metal–insulator transition should occur when $B/U = 1.15$. Hubbard's analysis did not include long-range interactions, and therefore did not predict any discontinuity in the number of current carriers.

In this book we treat the discontinuous nature of the transition using an analysis introduced by Brinkman and Rice (1970a, b). This applies to band-crossing transitions and transitions in an array of one-electron centres. We term the latter "Mott transitions" when the centres have a moment; we do not limit the term to cases when the moment is that of a single spin, and indeed such cases are rare (Chapter 3). The insulating antiferromagnetic state is sometimes called a "Mott insulator". A Mott transition can be accompanied by a change of structure (see Section 3 below).

2 Band-crossing transitions

These have been considered in Chapter 1, Section 3 in the approximation of non-interacting electrons. Figure 4.1(a) shows schematically the band structure of a non-metal with an indirect gap; we suppose that, through a change of volume or (in an alloy) a change of composition, the gap decreases and eventually becomes negative, as in Fig. 4.1(b). The system will then become metallic; there will be a condensed gas of electrons in the conduction band and of holes in the valence band; the Fermi energy E_F will be as shown. We shall now prove that the transition from insulator to metal is discontinuous, the number of carriers jumping from zero to a critical value n_c. The proof depends on the analysis given to explain the phenomenon of electron–hole droplets.† When crystalline germanium, for example, is strongly illuminated, or electrons and holes are introduced by double injection, the electrons and holes condense into droplets, giving through their recombination a photoluminescence line of frequency that does not depend on the intensity of illumination, unless the excitation is so strong that the droplets expand to fill the whole specimen (Nakamura and Morigaki 1974). The droplets form because the electron–hole gas has a minimum energy for a certain concentration n_c of electron–hole pairs. For isotropic energy surfaces, neglecting electron–electron and hole–hole correlation, and assuming a constant density of electrons and holes, an estimate of n_c can be obtained as follows. Writing

$$m^* = \frac{m_e m_h}{m_e + m_h},$$

where m_e and m_h are the effective masses of electrons and holes, the kinetic energy of the condensed gases at zero temperature is

$$\frac{3}{5}\frac{\hbar^2 n^{2/3}}{2m^*}$$

and their mutual potential energy is

$$-1.1e^2 n^{1/3}/\kappa,$$

where κ is the background dielectric constant. The minimum energy occurs when $n = n_c$, where

$$n_c^{1/3} a_H = 0.1 \tag{2}$$

and $a_H = \hbar^2 \kappa / me^2$. The energy of the electron–hole gas, per pair, is E_{crit}, where

$$E_{crit} = 0.1 m^* e^4 / 2\hbar^2 \kappa^2. \tag{3}$$

This is a much smaller energy than that of a free exciton ($-m^* e^4/2\hbar^2\kappa^2$), and a condensed gas of excitons is a possibility that has been discussed in the literature

---

† For reviews see Rice (1977), Hensel et al. (1977) and Keldysh (1986). For the original work see Rogachev (1968).

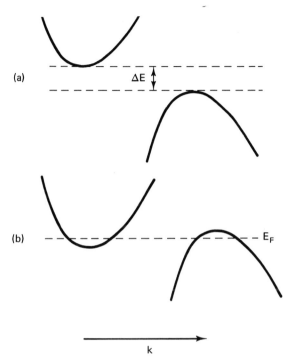

Fig. 4.1   Metal–insulator transition of band-crossing type. $E(k)$ is plotted against k in the insulating (a) and metallic (b) states.

(Keldysh and Kopaev 1965, Kohn 1967, Halperin and Rice 1968a, b).† However, it appears that when the surfaces are anisotropic (Brinkman and Rice 1973) and correlation is included (Vashishta *et al.* 1973), the reverse is usually the case.

The application to metal–insulator transitions is as follows. As the quantity ΔE in Fig. 4.1 decreases, and before it disappears but when $\Delta E = E_{\text{crit}}$, the number of carriers will, at zero temperature, increase from zero to a value of n_c given by (2). If κ is large, as it normally will be because the band gap is small, the discontinuity may be small and observable only at low temperatures.

In principle, then, the zero-temperature free energy of the system, plotted against volume or (in an alloy) composition, both denoted by x, must show a kink as illustrated in Fig. 4.2 at the metal–insulator transition. If x is the volume and this is decreased by pressure then there will be a discontinuous change of volume between B and A. If x denotes the composition then between B and A the alloy will be unstable, and will in equilibrium separate into two phases. The behaviour

---

† There is some evidence for them in band-crossing situations. Thus Mase and Sakai (1971) and Fukuyama and Nagai (1971a, b) obtained evidence of excitonic effects by shrinking the electron and hole orbits in bismuth in a magnetic field; an acoustic anomaly has been interpreted in this way. According to Tosatti and Anderson (1974), such effects are more likely in two dimensions than in three.

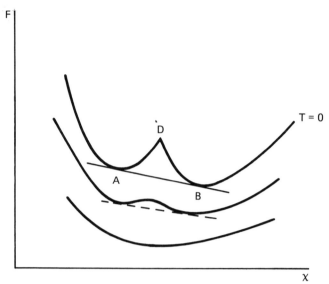

Fig. 4.2 Free energy F for various temperatures as a function of x near the metal–insulator transition of band-crossing or Mott type. x is volume or, in an alloy of type $M_{1-x}N_x$, composition. Between A and B the system is unstable.

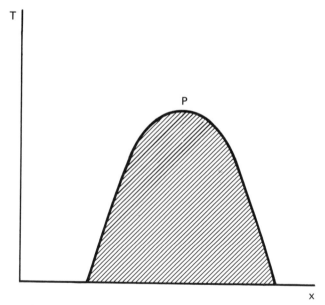

Fig. 4.3 The two-phase region (shaded) for the system illustrated in Fig. 4.2. P is the critical point.

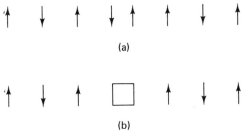

(a)

(b)

Fig. 4.4   The motion of (a) an electron in an upper Hubbard band and (b) a hole, marked by a square, in the lower band.

of the system at P, and the predicted discontinuous change in n, can only be observed in quenched alloys. Such a system in equilibrium will show a critical point as in Fig. 4.3.

These effects have not been observed in band-crossing transitions, but have in the so-called "Mott–Hubbard transitions" described in the next section.

3   Mott–Hubbard transitions

We consider here a crystalline array of one-electron centres, the electron being in a 1s-state. A doped semiconductor would be an example, if the donors were arranged in a crystalline array. Actually, there is no known crystalline system of this kind that undergoes a metal–insulator transition; we have to generalize our analysis for p- or d-states, and (apart from the high-T_c superconductors like La_2CuO_4) more than one electron or hole per centre, with the spins on a given atom coupled, according to Hund's rule, parallel to each other.

We suppose, as in Chapter 3, that where the distance between the atoms is large, the spins form an antiferromagnetic array, and that the system is insulating. If, however, an additional electron is introduced, it can sit on one of the atoms (labelled i) as shown in Fig. 4.4(a). We denote by ψ_i the many-electron wave function with an extra electron on atom i. Then a state in which the electron moves with wavenumber k can be described by the many-electron wave function

$$\sum_i e^{ika_i}\psi_i, \tag{4}$$

a_i denoting the lattice site i. These states form a band of energies—this is called the upper Hubbard band. We denote its width by B_1. In the same way, if, as in Fig. 4.4(b), an electron is removed from the atom i, and ψ_i' is the many-electron wave function for this system, then

$$\sum_i e^{ika_i}\psi_i' \tag{5}$$

is the wave function representing the movement of this "hole". Its band of energies is called the lower Hubbard band, and we denote its width by B_2.

If we neglect the electron–hole interaction, as in Chapter 1, then a metal–insulator transition should occur when the two bands overlap. For infinite values of a, the separation in energy between the two bands should be just the Hubbard U, so the transition occurs when

$$\tfrac{1}{2}(B_1 + B_2) = U. \tag{6}$$

The discontinuity should be treated just as for band-crossing transitions. The present author has suggested (see the first edition of this book) that it should be more marked for Mott transitions, because the dielectric constant κ for band-crossing will, as the forbidden gap diminishes, normally be large, while for Mott transitions the matrix elements determining κ will involve a movement of the electron from one centre to a neighbour, which should diminish the relevant oscillator strength. We shall show in Chapter 10 in a discussion of liquids that the two-phase region predicted in Fig. 4.3 is observed for fluid caesium and for metal–ammonia solutions (Mott transitions), but not for mercury, where the $6s^2$ state of the valence electrons indicates a transition of band-crossing type.

We turn now to an evaluation of n_c, the concentration of centres at which the transition occurs. We remark first of all that an experimental value is difficult to obtain. We do not know of a crystalline system, with one electron per centre in an s-state, that shows a Mott transition. Figure 5.3 in the next chapter shows the well-known plot given by Edwards and Sienko (1978) for n_c versus the hydrogen radius a_H for a large number of doped semiconductors, giving $n_c a_H = 0.26$. In all of these the positions of the donors are random, and it is now believed that for many, if not all, the transition is of Anderson type. In fluid caesium and metal–ammonia solutions the two-phase region is expected, but this is complicated by the tendency of one-electron centres to form diamagnetic pairs (as they do in VO_2). In the Mott transition in transitional-metal oxides the electrons are in d-states.

If we take $U = \tfrac{5}{8} e^2 / \kappa a_H$ and $B_1 = B_2 = 2zI$ with (Chapter 1, equation (11))

$$I = I_0 e^{-R/a_H},$$

where $R = n_c^{-1/3}$, then, with $z = 6$, we find

$$R/a_H \approx \ln 96 \approx 4.6,$$

whence

$$n_c^{1/3} a_H = 0.22, \tag{7}$$

which agrees well with the results of Fig. 5.3. Since the number appears as the reciprocal of a logarithm of a large number, changing (for instance) z from 6 to 12 will decrease n_c by only about 10%. We shall show in Chapter 5 that, for the same reason, the condition for the Anderson transition is similar: B has to be equated to a disorder parameter V_0, less than U, so that $n_c^{1/3} a_H$ is slightly smaller.

The assumption that $B_1 = B_2$ is of course incorrect, particularly for one-electron centres such as donors in semiconductors. Equally serious is the assumption that the bandwidth B can be calculated neglecting the antiferromagnetic structure, so that electrons can jump to next-nearest

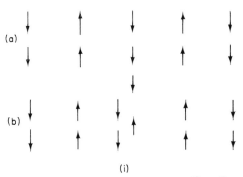

(i)

Fig. 4.5 Schematic representation of extra electrons in a d^2 antiferromagnetic insulator. (i) denotes the atom on which is the current carrier.

neighbours. This assumption seems at first sight wrong, because Fig. 4.4(a) suggests that an electron can jump only to a *next*-nearest neighbour, and I_1, I_2 should be the very much smaller overlap integrals between next-nearest neighbours. But this is not so, for the following reasons. If the atomic orbitals are degenerate, for example 3d-orbitals with two electrons per atom, the carrier will be coupled with its spin parallel to that of the moment of the atom on which it sits by the Hund's-rule energy (Fig. 4.5). One might think this energy would inhibit hops to a nearest neighbour, but this is not the case; the electron can hop to the state with antiparallel spins that has a symmetrical spin wave function and an antisymmetrical orbital function, which has the same energy. For non-degenerate states the hopping process shown in Fig. 4.5(b) has low probability, and occurs only because the nearest-neighbour states are hybridized with wave functions with antiparallel spin. The weight of these, as we have seen in Chapter 3, is

$$z\left(\frac{I}{U}\right)^2 = \frac{1}{4z}\left(\frac{B}{U}\right)^2.$$

Even when $B \approx U$, this is still not large, and if jumps to nearest neighbours were allowed only on account of this term then the mass of a carrier at the extremities of the Hubbard bands would be much enhanced. We think, however, that a more important effect allowing the carrier to move to nearest neighbours is the formation of spin polarons. These will be considered in Section 4.

In materials in which a metal–insulator transition takes place the antiferromagnetic insulating state is not the only non-metallic one possible. Thus in VO_2 and its alloys, which in the metallic state have the rutile structure, at low temperatures the vanadium atoms form pairs along the c-axis and the moments disappear. This gives the possibility of describing the low-temperature phase by normal band theory, but this would certainly be a bad approximation; the Hubbard U is still the major term in determining the band gap. One ought to describe each pair by a "London–Heitler" type of wave function

$$[\psi_a(1)\psi_b(2) + \psi_a(2)\psi_b(1)]\,[\chi_\uparrow(1)\chi_\downarrow(2) - \chi_\uparrow(2)\chi_\downarrow(1)]$$

and build up a Slater determinant from these functions.

A further non-metallic state without antiferromagnetic order is that with "resonating valence bonds", proposed by Anderson (1987). This has been described in Chapter 3.

Another way of reducing the correlation energy U that can occur for *degenerate* d-states in the following. If $\phi_i(x)$ and $\psi_i(x)$ are atomic orbitals corresponding to two degenerate d-states on the same site i then a superlattice of the form

$$\sum_{i=0,2,\ldots} e^{i\mathbf{k}\cdot\mathbf{a}_i}\,\phi_i + \sum_{i=1,3,\ldots} e^{i\mathbf{k}\cdot\mathbf{a}_i}\,\psi_i$$

can describe electrons with one spin, with a function obtained by reversing ϕ and ψ describing the other. This wave function gives no antiferromagnetic order, but is likely to distort the lattice, leading to Mössbauer splitting. The energy gained is the difference between U and the integral

$$\int \psi^*(1)\phi^*(1)\psi(2)\phi(2)\frac{e^2}{r_{12}}\,\mathrm{d}^3x_1\,\mathrm{d}^3x_2. \tag{8}$$

Examples of this behaviour may be $NbSe_2$ (Mott and Zinamon 1970) and TaS_2 (Williams *et al.* 1974).

4   Spin polarons and the Hubbard bands

Our main purpose in this section is to estimate the effective masses of carriers moving in the two Hubbard bands so that we can determine the transfer integrals I_1 and I_2 and hence the condition for the metal–non-metal transition. Hubbard bands, as we have stated, differ in an important way from the conduction bands of non-magnetic materials, because a carrier will tend to orient the moments of neighbouring atoms either parallel or antiparallel to its own spin. It thus forms round it a region in which the moments were disturbed, which it carries along with it when it moves. This region is called a "spin polaron", and its existence can have a large effect on the effective mass.

Spin polarons due to carriers in Hubbard bands differ in some ways from the spin polarons due to moments in f-states, etc., described in Chapter 3, Section 4 for EuSe and similar compounds, because the carrier and the moment on a given atom are in orbitals with the *same* quantum numbers (e.g. 3d). If the state of each atom is degenerate then the coupling is ferromagnetic, and the coupling constant J_H is the Hund's-rule coupling for two electrons in different 3d-states (Chapter 3, Section 1). If it is non-degenerate, however, two electrons with parallel spins cannot then go on to the same atom, and the coupling between the carrier and the moment is antiferromagnetic. It arises because the kinetic energy of the carrier is lowered if it can move from atom to atom. As in Chapter 3, Section 4, we suppose that if the radius of the polaron is R then the kinetic energy of the carrier in the "box" in which it can move freely is $\hbar\pi^2/2mR^2$. To this we must add the work required to line up ferromagnetically all the moments in a sphere of radius R; this is $\frac{1}{2}(\frac{4}{3}\pi)(R/a)^3 z I^2/U$. The radius of the polaron is thus obtained by minimizing the energy of the carrier and surrounding spins, namely

$$\frac{\hbar^2\pi^2}{2mR^2} + \frac{2\pi}{3}\left(\frac{R}{a}\right)^3\frac{zI^2}{U} - \tfrac{1}{2}B,$$

whence, as in Chapter 3, equation (12), we find

$$\frac{R}{a} = \left(\frac{\pi U\hbar^2}{2ma^2 zI^2}\right)^{1/5}, \tag{9}$$

and, since $I = \hbar^2/2ma^2$, this becomes

$$\frac{R}{a} = \left(\frac{\pi U}{zI}\right)^{1/5}. \tag{10}$$

As U/I (or U/B) goes to infinity, then so does R/a. Moreover, the energy of the spin polaron at rest tends to $-\tfrac{1}{2}B$. It follows that the bandwidth is B and, in the limit when U/B is large, is unaffected by the Hubbard U.

Our estimate of the band form should therefore show strong peaks at the extremities due to the (heavy) spin polarons. For these, k is a good quantum number, as long as $kR < 1$. If this is not so, the polaron concept breaks down and we come into the region investigated by Brinkman and Rice (1970a) where the electron loses energy rapidly to spin excitations and k is a band quantum number. These authors estimate that the bandwidth contracts by about 70%.

We are interested in the situation near the metal–insulator transition when $U \approx 2zI$, and for this case, from (10),

$$R/a \approx (2\pi)^{1/5} = 1.4.$$

Our conclusion is, then, that near the transition the spin polaron spreads only to the nearest and perhaps next-nearest neighbours.

Such spin polarons should not have a mass much greater than m in Si:P. Moreover, they can pass freely from one atom to another, and are *not* impeded by the antiferromagnetic order. Thus the bandwidth of each Hubbard band should, we believe, still be of order $2zI$, as it is for large values of U/B, and the equation

$$U = 2zI = B \tag{11}$$

should give a fair estimate for the metal–insulator transition, if we write $B = \tfrac{1}{2}(B_1 + B_2)$.

5 A degenerate gas of spin polarons; antiferromagnetic metals

If the spin-polaron model is correct, we must describe the carriers in the antiferromagnetic semimetal formed when the two Hubbard bands overlap as a *degenerate* gas of spin polarons; it should have the following properties.

(a) It is a semimetal, current being carried by electrons and holes.

(b) If η is the number of electrons and of holes per atom, the Fermi energies will be of order $E_F \sim \hbar^2 \pi^2 \eta^{2/3}/ma^2$. The Pauli susceptibility χ will be enhanced, because the elementary magnets are spin polarons, each one having a large moment. If v is the number of moments in each polaron then χ will be given by

$$\chi \sim N\eta(\mu v)^2/E_F. \tag{12}$$

v cannot be greater than $(2\eta)^{-1}$, so the factor v^2 can lead to a decrease in χ as ξ increases. If v has its maximum value, the enhancement factor for the Pauli susceptibility is thus $\eta^{-5/3}$. The theory of Brinkman and Rice (1970b), to be discussed in Section 6, shows that when the antiferromagnetic order has disappeared the enhancement factor for the electronic specific heat is about the same as for the susceptibility. Whether this is so for antiferromagnetic metals needs to be proved. Experimentally, it seems to be the case for the antiferromagnetic metal V_7O_{13} (McWhan et al. 1973), for which the enhancement is about the same for χ and γ, and γ is as large as for $(V_{0.9}Ti_{0.1})_2O_3$, a non-magnetic "metal" with strongly enhanced density of states (see Chapter 6).

A point of some interest is the value of the moment on each atom and the Néel temperature at the metal–insulator transition. Putting $B=U$ in equation (6) of Chapter 3 shows that the moment is reduced by a factor of $1-(4z)^{-1}$ and that the Néel temperature at the transition, since $B=U$, is given by

$$k_B T_N = B/2z. \tag{13}$$

In an ideal case we should expect very high Néel temperatures; since for transitional-metal oxides $B \sim 1$ eV, if $2z = 10$ then we expect $T_N \sim 1200$ K. But in transitional-metal compounds, as we shall see in Chapter 6, the equation that we must use to determine the condition for the transition is

$$U_{scr} = B, \tag{14}$$

where U_{scr} is a value reduced by screening, i.e. by the polarization of the medium round the two charges, and $U_{scr} \ll U$. For the Néel temperature we write $k_B T_N \approx B^2/2zU$, so that

$$k_B T_N \approx \frac{B}{2z}\frac{U_{scr}}{U},$$

which may be of order $0.01B$, giving $T_N \sim 100$ K.

In real materials the quantity that we have called U/B can be varied in several ways. A decrease in volume will normally increase B. Changing c/a, by altering the overlap between bands, can also affect both parameters (cf. the discussion of V_2O_3 in Chapter 6). Alloying can also change the ratio.

The description given here is appropriate to the case when the atomic orbital is non-degenerate (e.g. 1s-states, or the d_{z^2} states in the rutile or carborundum

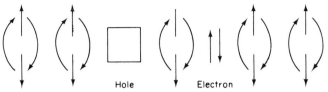

Fig. 4.6 The highly correlated electron gas of Brinkman and Rice. The doubly occupied and unoccupied states are shown. On the singly occupied sites the moment can change direction each time an electron or hole passes (Mott 1972b).

structures). It is also appropriate to (say) t_{2g} d-states occupied by three electrons in a high-spin configuration. If, however, the overlapping electron can move into a state with spin *parallel* to the moments then the spin polarons formed will have moments *parallel* to that of the carrier. This is similar to the case discussed in Chapter 3, Section 4. The coupling is then due to the Hund's-rule term J_H (Chapter 3, Section 1). We think that there should then be a strong tendency to ferromagnetism (cf. the discussion of CrO_2 in Chapter 6).

6 Disappearance of the moments; the Brinkman–Rice highly correlated gas

In the antiferromagnetic metal the mean value $\langle \sigma_z \rangle$ of the spin on any atom can be less than $\frac{1}{2}\hbar$, because the spin is resonating between its two directions. As η (the number of carriers) increases, $\langle \sigma_z \rangle$ should decrease, according to the argument of Chapter 3, Sections 7 and 10, as $(x - x_0)^{1/2}$, where x is any parameter that alters B/U. We estimate the value x_0 where the moments disappear as that at which the volume of the spin polaron is equal to the mean volume per carrier; the moment is then always more strongly affected by the moving carrier than by the antiferromagnetic interaction I^2/U. Assuming that a spin polaron occupies a given atom and its nearest neighbours, we see that the moments should disappear when

$$\eta \approx \frac{1}{2(z+1)}. \tag{15}$$

It need hardly be stressed that this estimate is approximate in the extreme.

The electron gas thus obtained, though "nearly antiferromagnetic", has no antiferromagnetic order. It is called "highly correlated" because only some 10% or less of the atomic sites contain two (or no) electrons; the others each contain one electron, of which the spin is resonating between the two possible positions. This behaviour is illustrated in Fig. 4.6.

The properties of such a gas are as follows:

(a) it must (by Luttinger's theorem, Chapter 2, Section 5) have a spherical Fermi surface, or a surface modified by the crystal structure but *not* split into two Hubbard bands;

(b) the Pauli susceptibility *and* the electronic specific heat are both greatly enhanced (contrast the nearly ferromagnetic gas of Chapter 3, Section 11, where only χ is enhanced);

(c) measurements of the Hall coefficient will give the total number of electrons.

(d) the number of current-carrying excitations is small.

Brinkman and Rice (1970b) were the first to show that the highly correlated electron gas has these properties. They started with the Hamiltonian

$$\sum_k \varepsilon_k (a_{k\uparrow}{}^+ a_{k\downarrow} + a_{k\downarrow}{}^+ a_{k\uparrow}) + U \sum_g a_{g\uparrow} a_{g\downarrow}{}^+ a_{g\downarrow} a_{g\uparrow}, \qquad (16)$$

where ^+a_k and ^+a_g are respectively the creation operators for electrons in a Bloch state k and a Wannier state g. They used a method due to Gutzwiller (1963, 1964), by which the energy is minimized as a function of η, the fraction of doubly occupied states. They found

$$\eta = \frac{1}{4} \left(1 - \frac{U}{C_0} \right). \qquad (17)$$

Here C_0 is eight times the modulus of the average energy without correlation, which will be perhaps about $\frac{1}{6}B$, depending on the band form; so

$$\eta \approx \frac{1}{4} \left(1 - \frac{\beta U}{B} \right),$$

where β is not far from unity. Thus, even if no antiferromagnetic state were formed, the number of doubly occupied states would become zero and the material would become non-conducting (but not antiferromagnetic) for some value of U/B not far from unity. Such a state does not in fact exist, since antiferromagnetic order, as discussed in Section 5, will be set up first. The important point about this analysis is that it reconciles a spherical Fermi surface that includes one electron per atom with a number of current carriers that, according to (15), can be quite small. Brinkman and Rice also predicted a mass enhancement due to correlation of

$$\frac{m^*}{m} = \frac{1}{1 - (U/C)^2} \approx \frac{1}{2\eta}, \qquad (18)$$

which determines the enhancement of the electronic specific heat and the Pauli susceptibility.†

This result can be obtained most simply in the following way. Suppose a field F acts for time δt small compared with the time of relaxation. The current δj produced by $N\eta$ current carriers of each sign is given by

$$\delta j = 2\eta N e^2 F \, \delta t / m.$$

---

† For degenerate states see Chao and Gutzwiller (1971) and Chao (1971).

Starting, however, with the spherical Fermi surface and effective mass m^*, the current is

$$\delta j = Ne^2 F \, \delta t / m^*.$$

Equating these two, we find

$$\frac{m^*}{m} = \frac{1}{2\eta},$$

(19)

which is Brinkman and Rice's result. They found that the Pauli susceptibility, which is inversely proportional to the effective mass, is enhanced by approximately the same factor. Formula (19) was used in Chapter 3, Section 8 to describe the Kondo mass enhancement.

This kind of enhancement, which occurs for a half-full band near the point where an antiferromagnetic lattice forms, is quite different from the Stoner enhancement for nearly ferromagnetic metals described in Chapter 3, Section 11. The latter occurs for non-integral occupation of a d-band, and enhances the Pauli susceptibility only, not the specific heat, apart from probably small paramagnon effects.

In some transitional-metal oxides discussed in Chapter 6, very large enhancement factors of order 70 are found. If our estimate that the material becomes antiferromagnetic when $\eta \approx [2(z+1)]^{-1}$ is anywhere near correct, the maximum enhancement should be $z+1$, or about 10. A possible way out of the difficulty is to assume that there is some dielectric polaron mass enhancement of the carriers, and in some materials (e.g. VO_x) this is probably so (Chapter 7, Section 3). However, we suggest that in addition an improvement of the Brinkman–Rice analysis is necessary along the following lines, which will give a somewhat larger enhancement. In their model there is no consideration of the correlation between the direction of the spin of the carrier and the direction of the surrounding spins. It pictures the carrier as jumping from site to site, the site having an even chance of changing its spin direction each time this happens. We think that an improvement would be to assign a definite spin to the carrier and allow it to make a "spin polaron" from the surrounding moments. This would lower the total energy. These spin polarons are to be treated as pseudoparticles forming two degenerate gases, one consisting of "electrons" and one of "holes". For these pseudoparticles we can introduce a density of states proportional to $E^{1/2}$, though broadened at the bottom owing to their finite lifetime. Each pseudoparticle carries a moment $\mu/2\eta$, so the susceptibility per electron is

$$\chi = \frac{2\eta}{E_F'} \left(\frac{\mu}{2\eta}\right)^2 = \frac{\mu^2}{2\eta E_F'}.$$

(20)

The Brinkman–Rice mass enhancement $(2\xi)^{-1}$ thus occurs, but E_F' is the Fermi energy for the current carriers, which is proportional (in a crystalline lattice) to $\eta^{2/3}$ and may be written $\eta^{2/3} E_F$. So an enhancement by $(2\eta^{5/3})^{-1}$ is expected for the susceptibility, which could be of order 50, nearer to the large values observed.

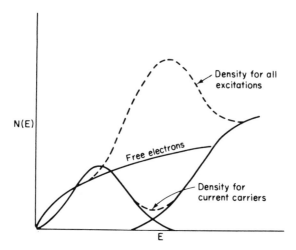

Fig. 4.7   Total density of states and density of current carriers in the highly correlated electron gas of Brinkman and Rice (Mott 1971).

Our model for the density of states is thus as in Fig. 4.7. The total density of states is mainly due to spin fluctuations, and has a maximum for $n = 1$, where n is the number of electrons per atom. The curve for current carriers needs to be used for calculating thermopower and resistance; the experimental evidence discussed in the following chapters suggests, however, that the Hall coefficient R_{H} is given by the classical formula $1/nec$.

If the sites are degenerate then the carrier and the moments of the spin polarons will be coupled *ferromagnetically*. The quantity J_{H} may be greater than I, especially if the latter is reduced by polaron formation. We may then have a ferromagnetic metal (CrO_2) or a strongly enhanced paramagnetism ($LaNiO_3$) with *small* temperature variation, as will be shown in Section 9.

The mass enhancement predicted by Brinkman and Rice for an array of one-electron centres using the tight-binding model is not shown by an electron gas in the field of a uniform positive charge ("jellium"). The analysis given by Hedin and Lundqvist (1969) gives for $R/a_{\mathrm{H}} \sim 5$ an enhancement of only 10% or less, according to the approximation used. It is to be assumed that a larger enhancement will occur as we approach the Wigner transition around $R/a_{\mathrm{H}} \sim 10$.

7   Slater's band-theory treatment of Mott–Hubbard insulators

Slater (1951) was the first to suggest that the insulating properties of antiferromagnetic transitional-metal compounds could be explained by supposing that the band was split by the antiferromagnetic lattice. Figure 4.8 shows how this could occur for a two-dimensional simple cubic lattice. The ground state of any antiferromagnetic insulator can be described in this way,

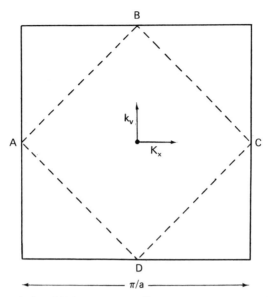

Fig. 4.8 Splitting of a band in k-space according to concepts proposed by Slater (1951).

though it must *not* be deduced that the Hubbard gap disappears above the Néel temperature, as we explain in Section 8. Here we confine our discussion to zero temperature. In Chapter 3, Section 2 we have described the electrons in antiferromagnetic insulators by Wannier localized functions. But it is well known that *for insulators* the Slater many-electron determinants using either Bloch or Wannier functions are identical. So Bloch functions can be formed from the Wannier functions for any insulating crystal. Thus it is legitimate to treat the ground state of these materials using an unrestricted Hartree–Fock approximation, in which the potential acting on the electron depends on its spin; and, indeed, in atoms with a moment antiparallel to the spin of the electron considered, the potential contains the term U, which is absent in the atoms with parallel moment because the electron does not interact with itself.†

Since we can use Bloch orbitals, and in an insulator the bands in this description are either full or empty, the crystal lattice together with any antiferromagnetic superlattice must be capable of splitting the bands so that this is possible; if this is not so then Luttinger's theorem (Chapter 2, Section 5) shows that the material must be metallic.‡ For antiferromagnetism in NiO or ferromagnetism in $CrCl_3$ this presents no difficulty. In the former the t_{2g} band with six electrons per atom is full and the e_g band, with room for four, is split into two sub-bands by the antiferromagnetic lattice, the energies of the two bands

---

† Some treatments, such as that of Wilson (1969, 1970), have taken the Hund's-rule term J_H. We do not believe this to be correct (cf. Mott and Zinamon 1970).
‡ This has of course been put in doubt by Anderson's proposal of resonating valence bond states (see Chapter 9).

being separated by a term of order U; one is occupied and the other empty. In ferromagnetic $CrCl_3$ the up-spin t_{2g} band will be fully occupied by the three electrons of Cr^{3+}, and the down-spin band will be empty. In a ferromagnetic non-metallic oxide with less than three electrons available this would not be possible. Another example is antiferromagnetic CoO, where the split t_{2g} and e_g bands cannot possibly provide a full band for seven electrons. In these materials the orbital moment is not quenched by the crystal field. The Slater principle suggests that there must be a more complicated superlattice of spins than in NiO, and indeed the work of Kahn and Erickson (1970) gives evidence that this is so (Chapter 3, Section 3).

One can see in an elementary way that a simple NiO-type antiferromagnetic arrangement cannot be stable in CoO. As in Chapter 3, Section 3, if $\phi_i(x)$ is the atomic orbital on atom i, we set for the Wannier wave function giving some overlap onto neighbouring atoms

$$\phi_i + A \sum_j \phi'_j,$$

where the summation is over the z neighbours and A is chosen to minimize the energy. But since the atomic orbit is degenerate, ϕ' need not—and in general will not—be the same orbital as ϕ; choosing a different orbital will increase (algebraically) the term $-2zIA$, but decrease the term UA^2. We can therefore see how the energy can be decreased if a superlattice of different orbitals is set up. Since spin–orbit coupling is present, the ϕs will be complex, giving some orbital moment, and the resultant of spin and orbit will show a more complicated superlattice. To the best of our knowledge, however, no calculations exist along these lines.

8   A generalization of the concept of a Mott transition

In any real material the specific volume will certainly change in an alloy at the composition for the transition—and frequently so will the crystal structure. The same is true if the transition takes place with increasing temperature, as a result of the decrease in the free energy of an electron near the Fermi surface together with soft phonon modes. In the latter case hysteresis may be observed at the transition, as in V_2O_3 (Fig. 6.2). It is sometimes argued (Honig 1985) that if hysteresis is observed, a transition should not be described as being of "Mott" type, since the latter is purely electronic. This is a matter of definition; we describe a Mott insulator as an antiferromagnetic material that would be a metal if no moments were formed. Probably this is true of all antiferromagnetic insulators, for instance compounds of the rare earths; without moments we might expect a partially filled 4f-band; but the material is far from any transition and its discussion is hardly useful in this case. The Mott transition, then, is a transition in such a material from a non-metallic state to a metal, which can be antiferromagnetic, and may have a different structure or specific volume.

We emphasize that, in the insulating state, the gap is given by $U - \frac{1}{2}(B_1 + B_2)$, and that it depends on the existence of moments and not on whether or not they are ordered. The gap is not related to the crystal structure. Indeed, if the crystal structure is such as to predict a gap, no antiferromagnetic lattice can form, unless the gap resulting from the Hubbard U is greater than that derived from the crystal structure.

9 Effect of temperature on antiferromagnetic insulators and highly correlated metals

For an antiferromagnetic insulator the most important property is that the magnitude of the Hubbard gap is not greatly affected when the temperature goes through the Néel point T_N (at least if $U/B \gg 1$). The Hubbard gap is then, essentially, the energy required to take an electron from one atom and to put it on a distant atom where another electron is already present. The energy U needed to achieve this is not dependent on the existence of the antiferromagnetic long-range order, and only small corrections are to be expected above the Néel temperature. The considerations of Section 7, according to which at zero temperature an insulator can be described in terms of the full and empty bands determined by long-range antiferromagnetic order, clearly cannot apply to the situation at a finite temperature. This situation is discussed in various theoretical papers; thus Langer et al. (1969) found for a simple cubic lattice of one-electron centres that if $U/B > 0.27$, a sharp metal–insulator transition occurs above the Néel temperature. Kimball and Schrieffer (1972), however, found that above the Néel temperature, for a rigid lattice, the transition between an antiferromagnetic insulator and a metal, as U/B varies, is not sharp, and we think that this result is probably correct if we start from the Hubbard model, even if long-range forces are taken into account, as we shall see in Section 10.

Ramirez et al. (1970) discussed a metal–insulator transition as the temperature rises, which is first order with no crystal distortion. The essence of the model is—in our terminology—that a lower Hubbard band (or localized states) lies just below a conduction band. Then, as electrons are excited into the conduction band, their coupling with the moments lowers the Néel temperature. Thus the disordering of the spins with consequent increase of entropy is accelerated. Ramirez et al. showed that a first-order transition to a *degenerate* gas in the conduction band, together with disordering of the moments, is possible. The entropy comes from the random direction of the moments, and the random positions of such atoms as have lost an electron. The results of Menth et al. (1969) on the conductivity of SmB_6 are discussed in these terms.

However, we do not think that this model is applicable to V_2O_3 and its alloys, chiefly because we do not think a conduction band necessarily lies near the lower Hubbard band, and it does not describe many important features, particularly the large mass enhancement (approximately 50) in the metallic phase. In our view, to understand the behaviour of such materials as the temperature is raised, and in particular the metal–insulator transition, we need to study the properties and especially the entropy of the highly correlated electron gas or of the antiferromagnetic metal at high temperatures. We shall take then as our model N one-electron atoms, which may be in degenerate (d^1) or non-degenerate (s^1) states, at finite temperature. We first consider non-degenerate states. The unenhanced bandwidth

$$2zI = z\hbar^2/ma^2$$

defines an effective mass probably not far from m_e. The Fermi energy E_F should be half this, $E_F = zI$, and it is this that will be determined by experiments on optical absorption or photoemission. Free-carrier absorption would determine this mass only at frequencies such that $\hbar\omega > zI$; at low frequencies the enhanced mass $m^* = m/2\xi$ would be determined. The high specific heat, enhanced by a factor $(2\eta)^{-1}$ or more, should lead to a *low* degeneracy temperature T_1, where

$$k_B T_1 = 2\eta E_F'. \tag{21}$$

Since the strength of the coupling of the moments to the carriers is about $\eta\hbar^2/ma^2$, we deduce that, for a temperature above T_1, the moments are free, giving entropy $Nk_B \ln 2$. If we think in terms of spin polarons, the polarons have broken up at this temperature. The excitations are mainly magnetic. The "carriers" will move through an array of disordered spins, and should behave like a non-degenerate gas with entropy

$$2N\eta k_B \ln\left(\frac{1}{\eta} - 1\right). \tag{22}$$

The formation of more carriers will involve energies of order zI, so we expect two degeneracy temperatures, T_1 and T_2, where T_2 is given by

$$T_2 = E_F/k_B.$$

The susceptibility is expected to appear as in Fig. 4.9. In the first edition of this book we quoted results due to Quirt and Marko (1973) on "metallic" Si:P near the concentration for the metal–insulator transition that show just this behaviour. However, it is now clear (see Chapter 5) that the transition in Si:P is not of "Mott" type, so this must be accidental.

If we are right in thinking that E_F' for the spin polarons is less than E_F/η then the polaron gas should become non-degenerate before they break up. We doubt if this will have any major effect on the susceptibility curve. However, if the orbitals are degenerate (e.g. d^1 states), and if the Hund's-rule coupling J_H is greater than I, the ferromagnetic spin polarons can survive above the temperature at which the carriers become non-degenerate. A relevant example is provided by the magnetic susceptibilities of metallic $LaNiO_3$ and $LaCuO_3$, in both of which the transitional-metal ion is *trivalent*, studied by Parent (1972). His results are shown in Fig. 4.10. For $LaCuO_3$ two electrons per Cu atom occupy half the states in an e_g band; we have here, then, a simple case of the Brinkman–Rice highly degenerate gas (like V_2O_3, VO, metallic VO_2, etc.) showing the large

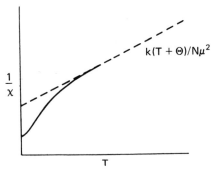

Fig. 4.9 Expected susceptibility as a function of temperature for a highly correlated electron gas. From Mott (1972 b).

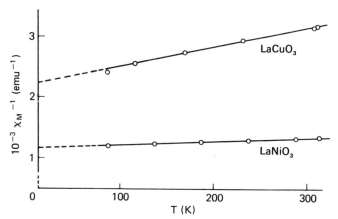

Fig. 4.10 Susceptibilities of $LaCuO_3$ and $LaNiO_3$ as functions of temperature. From Parent (1972); see also Goodenough *et al.* (1973).

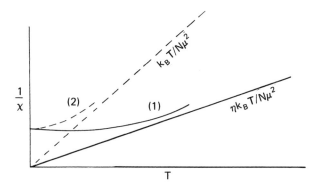

Fig. 4.11 (1) Expected susceptibility of a correlated electron gas when the atomic orbital is degenerate. (2) Behaviour of uncorrelated gas.

temperature dependence of χ already discussed. For $LaNiO_3$ there is *one* electron in the e_g band, allowing ferromagnetic spin polarons. It will be seen that the susceptibility is higher but the temperature dependence less than for $LaCuO_3$. This we believe is evidence that the highly correlated gas may be considered as being made up of spin polarons coupled ferromagnetically by Hund's rule, and that they can become non-degenerate *without breaking up*. Thus at moderately high temperature the susceptibility will behave as $N\eta(\mu/\eta)^2 k_B T$. Figure 4.11 shows the expected behaviour. This interpretation is due to Goodenough *et al.* (1973), who also suggested that, while $LaCuO_3$ must be a typical highly correlated metal, $LaNiO_3$ may be describable through Stoner enhancement of the susceptibility, which increases χ but not its temperature dependence or the electronic specific heat (Chapter 3, Section 11).

We now suppose that the carriers can form dielectric polarons, with mass enhancement up to m_p, but no hopping. If so, our temperature T_1 becomes T'_1, where

$$k_B T_1 = 2\eta E_F m/m_p,$$

but E_F and T_2 are unaltered. This mass enhancement, assumed to be by a factor less than about 10, will hardly affect the capacity of the carriers to form spin polarons, since, according to (9), the radius of the spin polaron is proportional to $m^{-1/5}$. We therefore think that a mass enhancement, affecting the Pauli susceptibility, of order $(m_p/m)/2\eta$ is to be expected.

If dielectric polarons are formed then the argument of Chapter 2, Section 2 suggests that η^{-1} must be at least about 10. The model used here may therefore give one reason why antiferromagnetic metallic compounds are so rare; in transitional-metal oxides where dielectric polarons are formed, if two bands, Hubbard or otherwise, overlap, there will be at least 10% of carriers. This is normally enough to wipe out the long-range antiferromagnetic order. The antiferromagnetic metallic compounds are normally sulphides or selenides (Chapter 7), where the d-band is wider, and probably small polarons are *not* formed. There is thus a much better chance of a small overlap between two Hubbard bands, giving an antiferromagnetic metal.†

A theoretical description of the a.c. conductivity has not been given. At low frequencies one would expect free-carrier absorption corresponding to the enhanced effective mass. At a frequency ω comparable to $E_F/\hbar$, where E_F is the unenhanced Fermi energy, we should expect the mass to go over to the unenhanced value. An example of this behaviour, taken from the infrared reflectivity of the vanadium oxides, is shown in Figs. 6.8 and 6.9.

As regards the thermopower, we can use the normal metallic formula

$$S = \frac{\pi^2}{3} \frac{k_B^2 T}{e} \frac{d \ln \sigma}{dE}.$$

---

† Crystalline $Li(NH_3)_4$ is probably an antiferromagnetic metal near the point where moments vanish (Glaunsinger *et al.* 1972).

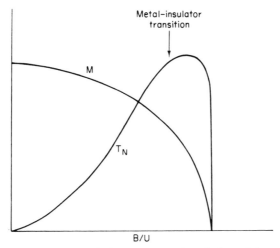

Fig. 4.12 Expected behaviour of the moment M and the Néel temperature T_N as functions of B/U.

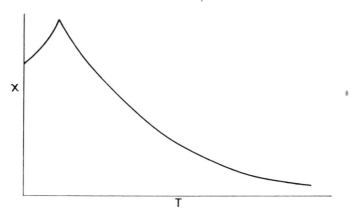

Fig. 4.13 Expected susceptibility as a function of temperature of a highly correlated antiferromagnetic electron gas near the value of B/U where moments disappear.

But, particularly if the mean free path is short, σ is a minimum for one electron per atom, so $d\sigma/dE$ may be small, and change sign if n, the number of electrons per atom, can be changed (as in VO_x, see Chapter 7, Section 3).

Next, we discuss the magnetic behaviour of the materials described here as B/U is varied. The moment at $T=0$ in the antiferromagnetic lattice and Goodenough's (1971) conjecture for the Néel temperature are shown in Fig. 4.12. There exists at present no good theoretical description of the way T_N drops as we go towards the value of B/U where the moments vanish. Near this value we do *not* expect an entropy $k_B \ln 2$ on going through the Néel point. With small values of $\langle \mu_z \rangle$ there is a quantum-mechanical resonance between the two directions of the moment, and the moments on a given atom are nearly equal, so $\langle \mu_z \rangle$ is small; at T_N they become equal. The susceptibility should appear as in Fig. 4.13.

10   Resistivity of highly correlated gas

The resistivity of a highly correlated gas in the absence of impurities or scattering by phonons has not been treated in detail. According to Monecke (1972), the conclusion of Chapter 2, Section 6 that for a *spherical* Fermi surface electron–electron (or paramagnon) scattering affects only the thermal but not the electrical resistance remains valid. For a non-spherical Fermi surface, however, a large term in the electrical resistance proportional to T^2 is expected. It is perhaps a matter of terminology whether we describe this as electron–electron (Landau–Baber) scattering or scattering by paramagnons (magnetic fluctuations); as we have seen, the excitations giving a large linear specific heat are in fact magnetic in origin. We expect then a resistance as shown in Fig. 2.9, flattening off at $T \sim T_1$, with T_1 given by (21). For the saturation value of the resistance the mean free path for the "carriers" is probably of the order of the lattice constant a, so

$$\sigma \approx S_F e^2 a / 12\pi^3 \hbar,$$

where S_F is the Fermi-surface area for 0.1 electrons per atom; σ is thus of order $1000 \, \Omega^{-1} \, \text{cm}^{-1}$. A curve of this kind for V_2O_3 is discussed in Chapter 6.

In the presence of impurities we think the resistivity should be calculated by taking the current carriers into account, and each impurity should give a cross-sectional area for them of order a^2. Thus the mean free path l is $1/N_0 a^2$, where N_0 is the concentration of impurities. The conductivity is given by

$$\sigma = \frac{2 S_F' e^2}{12 \pi^3 \hbar N_0 a^2},$$

where S_F' is the Fermi-surface area for current carriers, so that $S_F' = \eta^{2/3} S_F$; here S_F is the true Fermi-surface area. Thus

$$\sigma = \frac{2 \eta^{2/3} e^2 S_F}{12 \pi^3 \hbar N_0 a^2}.$$

The effective cross-section is enhanced by a factor $\frac{1}{2} \eta^{2/3}$. McWhan *et al.* (1973a) pointed out that in V_2O_3, dilute solutions of Cr_2O_3 give a cross-section greater than a^2 per Cr atom, which tends to confirm our analysis.

5

Interacting Electrons in Non-Crystalline Systems. Impurity Bands and Metal–Insulator Transitions in Doped Semiconductors

1 Introduction

In Chapter 1 we have outlined Anderson's (1958) theory of localization for a model of non-interacting electrons, defined a mobility edge E_c and shown how a degenerate electron gas can undergo a metal–insulator transition when the Fermi energy E_F passes through E_c, so that $E_F - E_c$ changes sign. In Chapter 2 we described some of the effects of electron–electron interaction on conduction in crystalline materials; for non-crystalline materials the theory, developed in the early part of the decade 1980–90 by scientists in the Soviet Union and elsewhere, shows that the effect of this interaction is more striking for a *degenerate* gas in a non-crystalline material than in crystals. Of course, for the electrons in the conduction band of a non-crystalline semiconductor, which form a non-degenerate gas, this interaction is too small to be significant. For these, on the other hand, the effect of phonons can be very important in delocalizing localized states, even at zero temperature; this, a subject lying outside the scope of this book, is reviewed briefly in Chapter 2, Section 2; for a full account see the book by Overhof and Thomas (1989), who initiated our understanding of this subject, Mott (1988) and Mott and Davis (1979).

For a condensed electron gas at zero temperature there is no such effect; near the Fermi energy localized states have a long lifetime for the same reason as that outlined in Chapter 2, Section 1; electrons cannot drop downwards in energy, and the Fermi energy remains sharp. Phonons at finite temperature have an important role, as in crystals, in giving rise to a scattering process, which—in contrast with scattering by impurities or by disorder—is inelastic. They can therefore be responsible at temperatures comparable to or well above the Debye temperature for an inelastic diffusion length L_i proportional to $T^{-1/2}$, although

145

at low temperatures electron–electron collisions have the major effect on L_i (Chapter 1, Section 10).

Interaction of electrons with phonons, and the fact that the presence of a trapped electron can deform the surrounding material, allows the radius ξ of an empty localized state to change when the state is occupied. Also, in a condensed electron gas phonons lead to a mass enhancement near the Fermi energy, or in some circumstances to polaron formation. For the development of the theory, and comparison with experiment, it is therefore desirable to begin by choosing a system where these effects are unimportant. The study of doped semiconductors provides such a system. This is because the radius a_H of a donor is given, apart from central cell corrections, by the hydrogen-like formula

$$a_H = \hbar^2 \kappa / m_e e^2, \tag{1}$$

where κ is the background dielectric constant and m_e the effective mass. In general $\kappa \gg 1$ and $m_e < m$, so a_H is considerably greater than the lattice constant, and therefore these effects are small. Systems in which they are important are discussed in later chapters, particularly Chapter 7.

In a doped n-type semiconductor electrons are bound in an array of hydrogen-like centres in random positions. The disorder in uncompensated samples comes from these random positions. For low concentrations these form a non-conducting system. At high concentrations there is a transition to a "metallic" state, in the sense that the conductivity extrapolated to zero temperature remains finite. Whether this is a "Mott" transition, depending on the Hubbard U, or one of Anderson type caused by disorder together with long-range interaction is uncertain and will be discussed in this chapter. In many-valley materials it appears to be of the latter type. But we emphasize here that we believe the former classification only to be meaningful within the tight-binding approximation, which is applicable to impurity bands, while long-range interaction gives effects observable in amorphous metals, and in any metallic material where the mean free path is short. However, the evidence is strong that in doped semiconductors the transition does in fact take place in an impurity band (Section 10).

Impurity conduction can also be studied in compensated semiconductors, i.e. materials containing acceptors as well as donors, the majority carriers (or the other way round). For such materials, even at low concentrations, activated hopping conduction can occur (Chapter 1, Section 15), some of the donors being unoccupied so that an electron can move from an occupied to an empty centre. Here too a metal–insulator transition can be observed, which is certainly of Anderson type, the insulating state being essentially a result of disorder.

2 Impurity conduction; metal–insulator transitions in impurity bands

The phenomenon now known as impurity conduction was first observed by Hung and Gleismann (1950) as a new conduction mechanism predominant at low

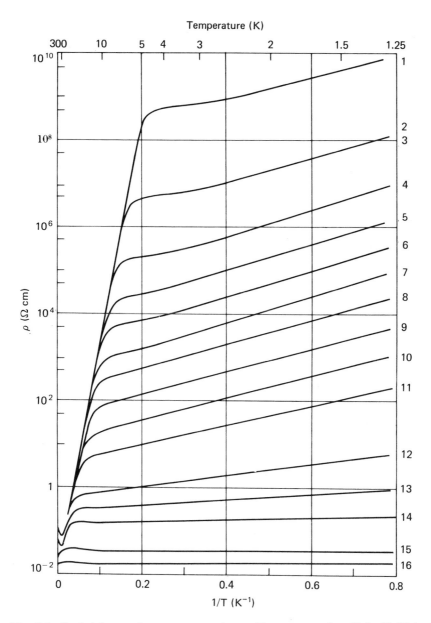

Fig. 5.1 Resistivity ρ of p-type germanium with compensation K $(=N_D/N_A)=0.4$ (Fritzsche and Cuevas 1960). The concentrations of acceptors are as follows (in cm^{-3}): (1) 7.5×10^{14}; (2) 1.4×10^{15}; (3) 1.5×10^{15}; (4) 2.7×10^{15}; (5) 3.6×10^{15}; (6) 4.9×10^{15}; (7) 7.2×10^{15}; (8) 9.0×10^{15}; (9) 1.4×10^{16}; (10) 2.4×10^{16}; (11) 3.5×10^{16}; (12) 7.3×10^{16}; (13) 1.0×10^{17}; (14) 1.5×10^{17}; (15) 8.8×10^{17}; (16) 1.35×10^{18}

temperatures in crystallline silicon and germanium that was doped and compensated. Some typical results due to Fritzsche and Cuevas (1960) are shown in Fig. 5.1. It will be seen that at low temperatures the conductivity behaves like

$$\sigma = \sigma_3 \exp\left(-\frac{\varepsilon_3}{k_B T}\right) \tag{2}$$

and that σ_3 depends strongly on concentration, except at high concentrations, where metallic behaviour occurs. This process, which we now call "nearest-neighbour hopping", was first described in detail by Kasuya and Koida (1958) and Miller and Abrahams (1960). Conwell (1956) and Mott (1956) first pointed out that this could only occur for compensated samples, and for low concentrations this was confirmed by the experimental work of Fritzsche (1958, 1959, 1960). Near the transition, as we shall see, this is not necessarily so. At sufficiently low temperatures (2) should always be replaced by an equation for variable-range hopping, such as

$$\sigma = \sigma_0 \exp\left[-\left(\frac{T_0}{T}\right)^\nu\right]. \tag{3}$$

Equation (3) was first proposed by Mott (1968) with $\nu = \frac{1}{4}$; Efros and Shklovskii (1975) showed that when electron–electron interaction is taken into account, in the limit of low T we expect $\nu = \frac{1}{2}$, though $\nu = \frac{1}{4}$ can be observed at higher temperatures, especially near the transition (see Chapter 1, Section 15).

We denote by ε_1 the energy for excitation of an electron into the conduction band (so that σ varies as $\exp(-\frac{1}{2}\varepsilon_1/k_B T)$ for an uncompensated semiconductor), and ε_3 is defined as above for nearest-neighbour hopping. Another form of conduction was identified by Fritzsche (1958), Fritzsche and Cuevas (1960) and Davis and Compton (1965). In this, as originally formulated, an electron is excited from a donor centre to an already-occupied centre, and then moves from one occupied centre to another. In the terminology of Chapter 4, it moves in the upper Hubbard band into which the impurity band is split. We now say that it is excited to the mobility edge in the upper Hubbard band, or, should the two bands have merged, to a mobility edge in the (merged) impurity band. We denote the excitation energy by ε_2. Figure 5.2 shows the results of Davis and Compton (1965) for ε_1, ε_2 and ε_3, as functions of the average distance a between donors in doped Ge.

It can be seen that ε_2 drops linearly at first, but has lower slope near the transition. There is no discontinuity, as would be expected for a Mott transition in a crystal (Chapter 4), and, as we believe, occurs (though broadened by temperature) in liquids such as fluid caesium (Chapter 10). The disorder here is greater than in a liquid metal because the orbitals of the electrons in the donors can overlap strongly. The present author (Mott 1978) has given conditions under which disorder can remove the discontinuity; but this may not be relevant to such materials as Si : P, because (Section 12) the Hubbard gap has disappeared, at any rate in many-valley semiconductors, at a concentration well below the transition,

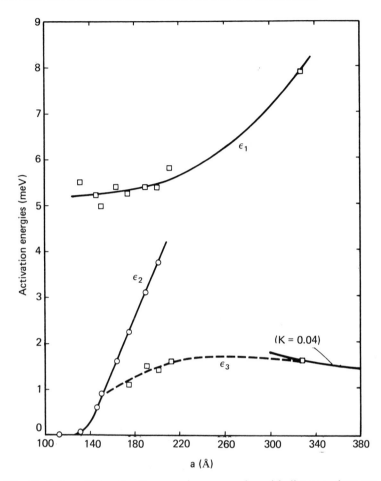

Fig. 5.2   Variation of the activation energies ε_1, ε_2 and ε_3 with distance a between donors for n-type germanium (Davis and Compton 1965). For ε_3 the calculations of Miller and Abrahams for $K = 0.04$ are shown (full curve).

and, as indicated in Section 1, the transition is of Anderson type, though strongly influenced by long-range interactions. This, we believe, is why Fig. 5.2 shows ε_2 tending to zero at $a \approx 140\,\text{Å}$, while the transition is at $a \approx 100\,\text{Å}$. Possibly the linear part of the plot corresponds to the range of a for which a Hubbard gap still exists, and the curved part—which should in principle be linear but may be curved in consequence of fluctuations in composition—has a smaller slope.

In the ε_2 regime the conductivity behaves like $\sigma_0 \exp(-\varepsilon_2/k_B T)$. As no electron–electron interaction is involved here, σ_0 should, as explained in Chapter 1, be equal to $0.03e^2/\hbar L_i$. The experimental evidence (Fritzsche 1978) is that $\sigma_0 = \sigma_{\min}$, where $\sigma_{\min} = 0.03e^2/\hbar a$. The present author (Mott 1987, p. 43, 1988) deduces that $L_i = a$, which means that every time an electron moves from one centre to another it loses energy to a phonon. In the above papers, by comparing the interaction with phonons leading to the mean free path in crystalline germanium, the present author finds this not improbable.

A preexponential factor equal to σ_{min} is a common feature of many systems, and must be ascribed in our view to strong interaction with phonons or spin waves. These systems are discussed further in Chapter 7.

In our discussion we assume that for concentrations up to and some way beyond that at which the transition occurs the electrons remain in an impurity band separate from the conduction band. At higher concentrations, perhaps $\gtrsim 3n_c$, the impurity band merges with the conduction band. This was first suggested by Alexander and Holcomb (1968). The evidence is summarized in Section 10.

3 Behaviour near the metal–insulator transition

Experimental investigations of the behaviour of doped semiconductors near the transition have shown the following.

(a) The value of the critical concentration n_c satisfies the equation

$$n_c^{1/3} a_H \approx 0.27 \tag{4}$$

for a very wide variety of (uncompensated) materials, as shown by Fig. 5.3, from Edwards and Sienko (1978), appropriate estimates of a_H being made. Equation (4) is consistent with a Mott transition (Chapter 4), but the constant for an Anderson transition, though leading to a slightly lower value of n_c, should not differ greatly.

(b) The conductivity σ extrapolated to zero temperature goes to zero normally as

$$\sigma \sim (n - n_c)^\nu,$$

with $\nu = 1$. There is no minimum metallic conductivity. Doped uncompensated silicon and germanium are exceptions, giving $\nu \approx \frac{1}{2}$. Compensated silicon shows values between $\frac{1}{2}$ and 1, tending towards 1 with increasing compensation (Thomas et al. 1987). $\nu = 1$ is found for many other systems, for instance a-Si$_{1-x}$Nb$_x$ (Hertel et al. 1983). For transitions caused by a magnetic field $\sigma \sim (H_c - H)^\nu$, with $\nu = 1$, is normally observed (see Section 9). An elegant method for determining ν was used by Katsumoto et al. (1988). These authors investigated the metal–insulator transition in (crystalline) Al$_{0.3}$Ga$_{0.7}$As, doped with deep-level silicon, by the method of persistent photoconductivity. They found that $\sigma = \sigma_0 [(n - n_c)/n_c]^\nu$, with $\nu = 1$, at temperatures in the millikelvin range. A magnetic field of order 0.5 T produced a transition to a metallic state, again as $(H_c - H)^\nu$, with $\nu = 1$. This is thought to be caused by the term $-1/L_H$ in equation (51) of Chapter 1 with L_i replaced by $L_H = (c\hbar/eH)^{1/2}$.

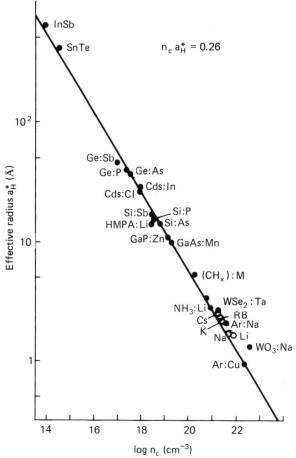

Fig. 5.3   Value of a_H plotted against $\log n_c$ at the transition (Edwards and Sienko 1978).

(c) The dielectric constant diverges on the insulating side with an index s such that

$$\kappa \sim (n_c - n)^s, \quad s = 2.$$

We believe that the localization length varies as

$$\xi \sim a \left(\frac{n_c}{n_c - n} \right)^\nu, \quad \nu = 1. \tag{5}$$

Assuming that $\sigma \propto e^2/\hbar\xi$ in the metallic regime, many observations support $\nu = 1$ both above and below the transition (cf. pp. 150, 157). On the other hand, calculations by Kramer and Mackinnon (cf. Chapter 1, Section 6.3) give $\nu = 1.6$, of course without electron–electron interactions. Mackinnon (private communication) suggests that interactions may result in $\nu = 1$. The only investigation that we know of for a non-degenerate gas (the conduction band of a-SiH) is that of Qui and Han (1988), which arrives at the result $v = 1.5$, giving some support to Mackinnon's hypothesis.

(d) The conductivity varies with temperature when $n > n_c$ as

$$\sigma = \sigma_0 + AT + mT^{1/2}, \tag{6}$$

or near the transition when σ_0 is small as $mT^{1/3}$. The term AT has already been discussed (Chapter 1, Section 9). The term proportional to $T^{1/2}$ has been observed in amorphous metals (Chapter 10) as well as in heavily doped semiconductors, and was first predicted by Altshuler and Aronov (1979). It is only significant when the mean free path is small, but it is *not* confined to the Ioffe–Regel regime ($l \sim a$). Thus in the semimetal bismuth this term, normally absent, appears as a result of cold-work, which decreases the mean free path but not to the limit $l \sim a$ (Bronovoĭ and Sharvin 1978, Bronovoi 1980). m is normally positive, so that σ increases with T, but it is negative for Si : P. It can change sign with magnetic field as it does in InP (Section 9).

4   Effect of interaction on the density of states and conductivity

The term $mT^{1/2}$ can be deduced from the influence of long-range electron–electron interaction on the density of states $N(E)$ near the Fermi energy. Altshuler and Aronov (1979) showed that

$$\delta N(E) = \frac{C}{2\pi^3 \hbar D l} \left[-1 + l \left(\frac{|E - E_F|}{\hbar D} \right)^{1/2} \right]. \tag{7}$$

The form of $N(E)$ is illustrated in Fig. 5.4. D is here the diffuison coefficient. Since, for l in the Ioffe–Regel regime, σ is proportional to g^2, on putting $|E - E_F| \sim k_B T$, we see that

$$\frac{\delta\sigma}{\sigma} = 2\frac{\delta N}{N} = \frac{C}{\pi^3 \hbar D l N(E_F)} \left(1 - \frac{l}{L_T} \right). \tag{8}$$

Here†

$$L_T = (\hbar D / k_B T)^{1/2}$$

is called the "interaction length". The Kawabata equation (Chapter 1, equation (52)) for a degenerate gas then becomes

$$\sigma = \sigma_B g^2 \left[1 - \frac{1}{(k_F l g)^2} \left(1 - \frac{l}{L_i} \right) - \frac{C}{(k_F l g)^2} \left(1 - \frac{l}{L_T} \right) \right], \tag{9}$$

---

† It will be noted that two electrons near the Fermi energy capable of being scattered will have energies differing by $\sim k_B T$. L_T is the length they diffuse before losing coherence (Dugdale 1987). We write this as $L_T = (D\tau_i)^{1/2}$. But, by the uncertainty principle, $\tau_i = \hbar / k_B T$, which leads to the above formula.

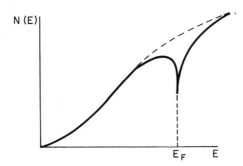

Fig. 5.4 Density of states in a non-crystalline material, showing the change near the Fermi energy E_F predicted by Altshuler and Aronov (1978).

where, if the interaction between electrons is of exchange type,

$$C = [\pi \hbar DlN(E)]^{-1}. \tag{10}$$

Taking for D and $N(E)$ their unperturbed values, we can see that the denominator $\pi \hbar DlN(E)$ in the last term is of order unity, so $C \sim 1$.

The constant C, however, depends on the nature of the interaction between a pair of electrons. If it is mainly of exchange type then C is positive and of order unity, as we have seen. There is also a term of "Hartree" type—namely that due to direct interaction between the electrons. This has the opposite sign. Finkelstein (1983) and Altshuler and Aronov (1983) found that when the Hartree term is included C should be multiplied by

$$\tfrac{4}{3} - \tfrac{2}{3}\tilde{F},$$

where

$$\tilde{F} = 32(1 + \tfrac{1}{2}F)^{3/2} - \frac{1 + \tfrac{3}{4}F}{3F} \tag{11}$$

and $F = x^{-1}\ln(1 + x)$, with $x = (2k_F\lambda_s)^2$, λ_s being a screening constant. Kaveh and Mott (1987) argue that modifications of this formula to take account of the many-valley nature of the conduction band in silicon and germanium are essential to explain a sufficiently large Hartree term to give a negative value† of C, so that m decreases with increasing concentration of donors (as in Si). The observations of Thomas et al. (1983) and Rosenbaum et al. (1983) show that m changes sign near the transition. This cannot, Kaveh and Mott argue, be ascribed to a drop in the Hartree term F; on the contrary, the screening length will decrease and F will increase as the transition is approached. They suggest (see also Kaveh et al. 1987) that L_i becomes smaller than L_T near the transition, so that L_i becomes dominant, and also (see Chapter 1) that instead of $\tau_i^{-1} \sim T^2$ we should write $\tau_i^{-1} \sim T^{3/2}$, so that $L_i \sim T^{-3/2}$ (Kaveh and Wiser 1984, p. 273). Thus the term due to quantum interference should tend to $T^{3/4}$.

---

† In Section 9 we find, however, that C is negative in InP.

The assumption that σ is proportional to the square of the density of states is normally deduced only in the regime $\sigma \lesssim \sigma_{min}$, and is based on the Edwards cancellation theorem. As shown in Chapter 1, Section 6.3, however, this is valid only if we can write $v_f = \hbar^{-1} dE/dk$ and deduce $v_f \propto g^{-1}$, which means a uniform expansion of the band. This is certainly not so for weak localization. For the singularity predicted by Altshuler and Aronov, any correction to dE/dk will be zero if averaged over a range $\sim k_B T$ about E_F. We consider then that if $\sigma > \sigma_{IR}$,

$$\frac{\delta\sigma}{\delta} = \frac{\delta N}{N},$$

without the factor of 2 in this range.

To summarize, then, we expect that normally C and m are positive, with both the quantum interference effect proportional to T (or eventually to $T^{3/4}$) and the many-body effect ($\propto T^{1/2}$) increasing the conductivity. Silicon and germanium are perhaps exceptions, as may be other many-valley materials. The change of sign observed in silicon near the transition is, as we have seen, due to the fact that in this regime the T dependence of σ is no longer caused by L_T.

5   The $T^{1/3}$ behaviour of the conductivity near the transition

In the case where C is positive, at the transition the conductivity should vary as $T^{1/3}$, and near the transition $\sigma = a + mT^{1/3}$ is observed. An example due to Maliepaard et al. (1988) is shown in Fig. 5.5.† The theoretical explanation, valid only for positive C in (9), is as follows. Near the transition, in the term

$$\frac{C}{(k_F ag)^2} \frac{a}{L},$$

which will determine the conductivity, L_T is equal to $(\hbar D/k_B T)^{1/2}$, but D is itself proportional to σ through the Einstein equation

$$\sigma = e^2 D N(E_F).$$

Thus we have

$$\sigma = \frac{e^2}{3\hbar a} \frac{C}{(k_F a)^2} a \left[\frac{k_B T N(E_F) e^2}{\sigma} \right]^{1/2}, \tag{12}$$

whence it is easily seen that $\sigma^{3/2}$ is proportional to $T^{1/2}$, so $\sigma \propto T^{1/3}$.

An interesting feature of Fig. 5.5 is that some specimens show metallic behaviour only above a certain temperature, and σ is then of the form $a + bT^{1/3}$, with a negative value of a. Below this temperature they are thought to be insulators, with conduction by hopping. We believe, following Thouless (1977), that when L_i or L_T become smaller than ξ localization can no longer occur and conduction will show metallic behaviour.

---

† Earlier results by Newson and Pepper (1986) for GsAs and InSb (n-type) show similar behaviour.

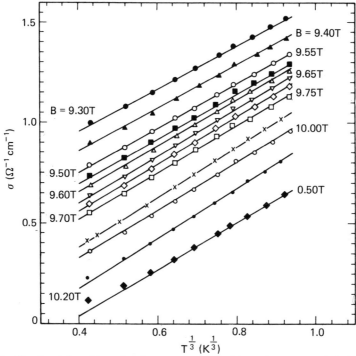

Fig. 5.5 Conductivity of doped and compensated GaS in various magnetic fields near the transition, plotted against $T^{1/3}$ (Maliepaard *et al.* 1988).

6 Deduction of the Altshuler–Aronov correction term in the density of states

In this section we deduce equation (7) for the correction to $N(E)$ resulting from long-range interaction, first predicted by Altshuler and Aronov (1979). The derivation is similar to that given by Mott and Kaveh (1985a). We have to consider the interaction $V(|\mathbf{r}_1 - \mathbf{r}_2|)$ between a pair of electrons. Taking only the exchange term, the resulting self-energy $S(E)$ for a given electron with energy E is

$$S(E) = -\sum_{E'} \int\int d^3r_1 \, d^3r_2 \, V(|\mathbf{r}_1 - \mathbf{r}_2|) \psi_E^*(\mathbf{r}_1) \psi_E^*(\mathbf{r}_2) \psi_{E'}(\mathbf{r}_1) \psi_{E'}(\mathbf{r}_2).$$

We write

$$V(|\mathbf{r}_1 - \mathbf{r}_2|) = \frac{1}{(2\pi)^3} \int d^3q \, V(q) \, e^{i\mathbf{q}\cdot(\mathbf{r}_1 - \mathbf{r}_2)}, \tag{13}$$

which defines $V(q)$. We then see that

$$S(E) = -\frac{1}{(2\pi)^3} \int dE' \, N(E') \int_{ql<1} d^3q \, V(q) |\langle \psi_E^* | e^{i\mathbf{q}\cdot\mathbf{r}} | \psi_{E'} \rangle|^2.$$

Using equation (67) of Chapter 1, we deduce that

$$S(E) = \frac{1}{8\pi^4 \hbar N(E_F)} \int dE' \, N(E') \int_{ql<1} d^3q \, V(q) \frac{Dq^2}{(Dq^2)^2 + (E-E')^2/\hbar^2}.$$

If the interaction $V(r)$ is screened according to the Thomas–Fermi law $V(r) = (e^2/\kappa r)e^{-r/\lambda}$ then

$$V(q) = \frac{1}{N(E_F)a^3},$$

the charge e cancelling out because $-\lambda^2 = \kappa/e^2 N(E_F)$. Thus the change in the density of states, given by

$$\delta N(E) = \frac{\partial S(E)}{\partial E}, \quad E = E_F,$$

becomes

$$\delta N(E) = \frac{1}{2\pi^3 \hbar Dl}\left[-1 + l\left(\frac{|E-E_F|}{\hbar D}\right)^{1/2}\right]. \tag{14}$$

This is (7), which is what we set out to prove in this section.

7   The metal–insulator transition in Si:P; the index $\nu = \frac{1}{2}$

The work of Thomas and co-workers on fine tuning of the transition by stress (Paalanen *et al.* 1982; for a review see Thomas 1985) showed that for Si:P at millikelvin temperatures the conductivity behaves like

$$\sigma = \sigma_0 (n - n_c)^\nu,$$

with $\nu \approx 0.5 \pm 0.1$. There have been many attempts to explain this behaviour. Paalanen *et al.* (1986), who measured the specific-heat anomaly near the transition and ascribe this to free spins, suggest that such spins may be relevant to this index. This has also been proposed by Kawabata (1984, 1988) and di Castro (1988). As far as we know, the only theory that gives a numerical value for the effect is that of Kaveh (1985a, b; see also Kaveh and Mott 1987), who suggests that the effect only occurs when C in (9) is positive, and thus normally only for many-valley materials, and that the correct value in this case is $\nu = \frac{1}{2}$. We now describe this theory. At zero temperature we can write

$$\sigma = \sigma_0\left[1 + \frac{C}{2\pi^3 N(E_F)\hbar Da} \frac{1}{(gk_F a)^2}\right],$$

and, substituting for D from the Einstein equation

$$\sigma = e^2 D N(E_F),$$

we find

$$\sigma = \sigma_0 \left(1 + \frac{C'e^2}{2\pi^3 ha\sigma} \right),$$

where $C' = C/(gk_F a)^2$ is positive. Near the transition, since $\sigma \to 0$, the second term is the larger. Therefore writing $\sigma_0 \approx 0.03 e^2/\hbar\xi$, we see that

$$\sigma^2 = 0.03 \frac{C'}{2\pi^3} \frac{e^2}{ha} \frac{e^2}{\hbar\xi}. \tag{15}$$

Since ξ behaves like $an_c/(n-n_c)$, it follows that

$$\sigma \sim C'^{1/2} \sigma_{\min} \left(\frac{n-n_c}{n_c} \right)^{1/2}. \tag{16}$$

$C'^{1/2}$ contains a factor $(gk_F a)^{-1}$, which should be about 1, and an unknown term from the Hartree interaction.

Experimentally, the work of Paalanen et al. (1982) gives

$$\sigma \approx 13\sigma_{\min} \left(\frac{n-n_c}{n_c} \right)^{1/2}, \tag{17}$$

so, if this theory is correct, the contribution of the Hartree term is suprisingly large. On the other hand, the Hartree term is expected to be reduced by a magnetic field (see also the discussion of InP in Section 9), so one would expect that, in a strong enough field, the index should tend to unity. Thomas et al. (1987) found that the zero-temperature conductivity of Ge : Sb does tend *linearly* to zero with increasing field, which in our view supports Kaveh's model.

A table showing experimental data is given by Milligan (1985); it shows that at the time of writing the index $v = \frac{1}{2}$ is found *only* in many-valley semiconductors in the absence of compensation. All other materials show $v = 1$.

Other theories of the effect (Kawabata 1984, Anderson 1985, Ruckenstein et al. 1985) all depend on the assumption of a large contribution to the resistivity coming from scattering by magnetic fields within the material. Fixed moments must certainly exist in such materials, but it seems to us that they would scatter electrons only through small angles and contribute little to the mean free path. The treatment by Kawabata (1984) explicitly supposes that the scattering by localized moments is strong. However, he does obtain good agreement with experiment for the product $\sigma_0 \kappa_0^{1/2}$, where κ_0 is defined in the next section.

8 The dielectric catastrophe

In Si : P and some other systems in which a metal–insulator transition takes place, such as the liquids K–KCl (see e.g. Warren 1985) and $Na_x NH_3$ (Chapter 10) the dielectric constant can become very large as the transition is approached from below, for example in Si : P by increasing the phosphorus content. Some

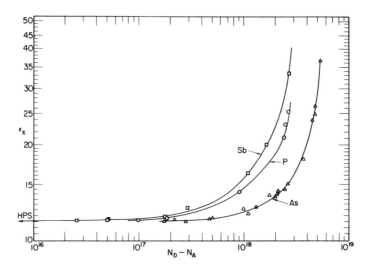

Fig. 5.6   Donor dielectric constant for Sb, P and As donors in silicon plotted against
concentration (Castner 1980).

experimental results for doped silicon are shown in Fig. 5.6. This can best be
discussed in terms of the Kramers–Kronig relationship

$$\kappa = \int \sigma(\omega)\frac{d\omega}{\omega^2}. \tag{18}$$

For $\sigma(\omega)$ we take the value given by the Kubo–Greenwood equation, namely
equation (34) of Chapter 1. Mott (1970) showed that $\sigma(\omega)$ is proportional to ω in
the localized regime, and Mott and Kaveh (1985b) give a corrected version

$$\sigma(\omega) = 4\pi^2 e^2 \hbar\omega^2 R_\omega^4 N(E_F)\xi, \tag{19}$$

where

$$R_\omega = \xi \ln\frac{H_0}{\hbar\omega},$$

and H_0 is the preexponential in the transfer integral between two sites a distance
r apart, namely $H_0 e^{-r/\xi}$. Equation (19) can be valid only up to a frequency ω_c
such that $\hbar\omega_c$ is the interval between sites in a volume ξ^3, so that

$$\hbar\omega_c = \frac{1}{\xi^3 N(E)}.$$

For $\omega > \omega_c$, we must use formulae of metallic types (Thouless 1977, Kawabata
1984), and, following Ortuno and Kaveh (1984), we write

$$\sigma(\omega) = 0.03 e^2/\hbar L_\omega, \tag{20}$$

where $L_\omega = (D/\omega)^{1/2}$, i.e. the distance an electron drifts in the period of the alternating field. Since $D = \sigma/N(E_F)e^2$, we find[†]

$$\sigma(\omega) = 0.03 \frac{e^2}{\hbar} [N(E_F)\hbar\omega]^{1/3}. \tag{21}$$

Turning now to the dielectric constant, using (18), we write

$$\kappa = \int_0^{\omega_c} \sigma_1(\omega) \frac{d\omega}{\omega^2} + \int_{\omega_c}^\infty \sigma_2(\omega) \frac{d\omega}{\omega^2},$$

where σ_1 and σ_2 are given by (19) and (21). The first term is

$$4\pi e^2 N(E)\xi^2$$

and the second is

$$\frac{2}{3} \omega_c^{-2/3} \frac{0.03 e^2}{\hbar} [\hbar N(E)]^{1/2}.$$

Both terms are proportional to ξ^2, so we expect κ to diverge with the same index as ξ^2, namely $v = 2$.

Paalanen et al. (1982), in their investigation of the dielectric constant κ in Si : P near the transition, find that

$$\kappa = \kappa_0 \left(\frac{n_c}{n_c - n} \right)^2.$$

In our view this suggests that the index for ξ is $v = 1$, unaltered by interaction, and whether or not quantum interference occurs.[‡] The considerations of Section 7 then suggest that $\sigma \sim \xi^{-1/2}$ for Si : P and other uncompensated many-valley materials, but for others $\sigma \sim \xi^{-1}$.

Other experimental evidence that the index for ξ is unity is scanty. If we are right in thinking that hopping conduction near the transition is not affected by a Coulomb gap (Chapter 1, Section 15), evidence can be obtained from this phenomenon; Castner's group (Shafarman et al. 1986) give evidence for $v = 1$ in Si : P. For early work in a two-dimensional system see Pollitt (1976) and Mott and Davis (1979, p. 138), which give evidence that $v = 1$. In two-dimensional systems there is evidence (Timp et al. 1986) that the Coulomb gap has little effect on hopping conduction.

Many authors have discussed a dielectric catastrophe and the Mott transition in terms of the Clausius–Mossotti relationship

$$\kappa = 1 + \frac{4\pi N\alpha}{1 - \frac{4}{3}\pi N\alpha}, \tag{22}$$

[†] The behaviour as $\omega^{1/3}$ is given by Götze (1979, 1981), Shapiro (1982) and Wölfle and Vollhardt (1982).

[‡] In Chapter 10 we suggest that it does not occur in liquids.

where α is the polarizability (Herzfeld 1927, Castner 1980, Edwards and Sienko 1983, Logan and Edwards 1985). In our view the role of (22) is the following. The equation is correct as long as the orbitals do not overlap appreciably, as first pointed out by Mott and Gurney (1940) in connection with the dielectric constants of alkali-halide crystals, which do not conform to (22). In fact, except for materials where cohesion is of the weak van der Waals type, overlap is probably too strong for (22) to be valid. On the other hand, if we consider the molecules of a solid at a distance a from each other then, for large a, (22) will be valid and will lead to an increase in κ. But this will in turn increase the atomic radius, and in non-crystalline materials it will also increase the localization length ξ, and thus promote overlap. In this way the divergence of the expression (22) may give a good indication of the concentration N for which the transition will occur, while the transition, when it happens, is of Anderson type. If this is so, an increase in ξ will also affect $\sigma(\omega)$.

In liquids, if L_i is small, it is possible that the whole rise in κ is due to (22).

9   Effect of a magnetic field

The main effects of a magnetic field are two-fold. A magnetic field can shrink the radius of a bound state, a process leading to a large positive magnetoresistance in the hopping regime (see e.g. Shklovskii and Efros 1984) and also, for conduction in an impurity band in the metallic regime due to reduction in the bandwidth B, leading to a metal-to-insulator transition. The other effect, in the metallic regime, comes from the modified Kawabata equation (cf. Chapter 1, Section 14)

$$\sigma = \sigma_B g^2 \left[1 - \frac{1}{(k_F l g)^2} \left(1 - \frac{l}{L} \right) \right],$$

where L in a magnetic field is given by

$$\frac{1}{L^4} = \frac{1}{L_i^4} + \frac{1}{L_H^4}$$

and where $L_H = (c\hbar/eH)^{1/2}$. If $L_H \ll L_i$, i.e. for strong fields, $L = L_H$, and this term leads to a negative magnetoresistance. Shapiro (1984) proposed that, on account of these two effects, the position of the mobility edge at zero temperature should show the behaviour sketched in Fig. 5.7; at low values of the field it moves to smaller energies, because the quantum interference effect is decreased, while for larger fields it shifts to higher energies. Thus if the Fermi energy for a degenerate gas lies at A, an insulator may make two transitions, one to metallic conduction and then one back to insulating. As we shall see, this effect has been observed in InP (Spriet et al. 1986) and in n-type GaAs (Maliepaard et al. 1989). If the Fermi energy lies at B, a transition to an insulating state must occur. We expect, in the neighbourhood of this transition, that

$$\sigma \sim (H_c - H)^\nu,\tag{23}$$

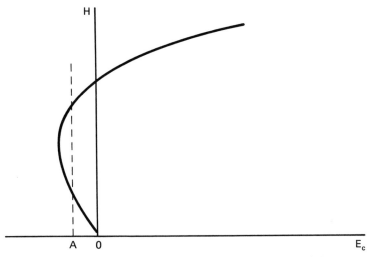

Fig. 5.7 Suggested position of the mobility edge E_c as a function of field H, as suggested by Shapiro (1984). If the mobility edge lies at A in the absence of a field, increasing the field will give *two* transitions.

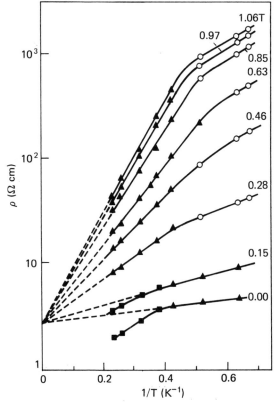

Fig. 5.8 Variation of the resistivity of a sample of InSb, doped and compensated for various fields (Ferré *et al.* 1975).

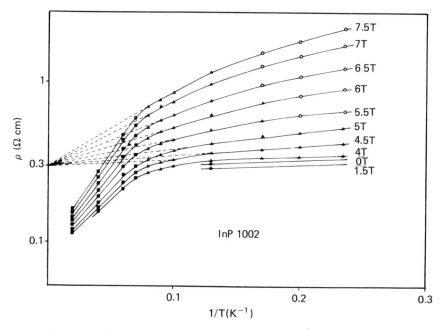

Fig. 5.9   Resistivity of n-type InP for varying magnetic fields (Biskupski 1982).

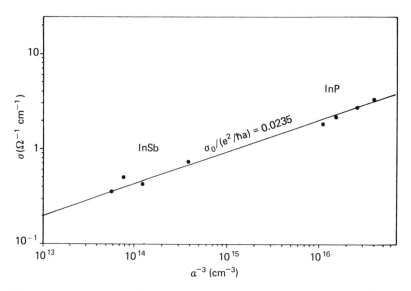

Fig. 5.10   Plot of σ_0 versus $a^{-3} = N_0 - N_A$, the reciprocal of the cube of distance between occupied donors, in InP and InSb (Biskupski 1982).

and the question arises as to what is the value of v? Hickami (1981) and others have predicted that $v = \frac{1}{2}$ for weak fields; but we do not know of any experimental work that establishes this, though the results of Spriet et al. (1986), reproduced in Fig. 5.9, show $v < 1$. If fields and temperatures are such that $H\mu_B > k_B T$, arguments are given by Imry and Gefen (1984) that $v = 1$, as a result of the suppression by the field of the Hartree term in the interaction. This increases the term C in (9), and if the increase is large enough then it will lead to a transition to an insulating state with $v = 1$. We think, however, that, even though the term $1 - a/L_H$ vanishes when $L_H = a$, the decrease in g and consequent increase of the term $-C(k_F l g)^2$ will eventually lead to a transition with $v = 1$ if C is positive; in our view, this should normally be the case in a strong magnetic field, which suppresses the Hartree term.

An index $v = 1$ in (23) is shown by GaAs (Maliepaard et al. 1989) and InSb (Mansfield et al. 1985).

The experimental work of Biskupski et al. (1984) and Long and Pepper (1984) on compensated samples of InP and InSb is of particular interest. Ferré et al. (1975) obtained the results shown in Fig. 5.8 for the resistivity of n-type InSb in the range of liquid-helium temperatures and fields up to 1.06 T. Later results for InP, obtained by Biskupski (1982), using stronger fields, are shown in Fig. 5.9. The preexponential factor σ_0 in $\sigma = \sigma_0 e^{-\varepsilon/k_B T}$ can be read off from the limit when $1/T = 0$, and is plotted against a^{-3} in Fig. 5.10, where $a = (N_D - N_A)^{-1/3}$ is the mean distance between *occupied* centres. σ_0 was found, for both InSb and InP, to be $0.0235 e^2/\hbar a$. As we showed earlier, we expect $0.03 e^2/\hbar L_i$ and must suppose, as for GeSb (Section 2), that $L_i \sim a$ in this work, on account of strong interaction with phonons.

Work in the millikelvin range revealed, however, a striking difference between InP and InSb. For InSb Mansfield et al. (1985) observed a disappearance of σ with $v = 1$ (not claimed by Biskupski et al. (1981), though this work was only at liquid-helium temperature). On the other hand, Biskupski et al. (1984) and Long and Pepper (1984) have both extended their work on InP to $T = 50$ mK and found an apparently discontinuous transition for a field of about 7 T, at which $\sigma = \sigma_{min}$. Mott (1989), examining the data given by Dubois et al. (1985), has come to the conclusion that this material is peculiar only in that the constant C in (9) is negative in the absence of a field, as is the case for many-valley semiconductors; this is certain because the $T^{1/2}$ term in σ, (6), is observed by these authors to be negative. As the field increases, C changes sign; it is zero for a field of about 7 T, for which σ is constant over a large range of T, and then drops to zero. The variation is not expected to be linear except near the transition, because C/g^2 behaves like a power of H.

The Hall effect in the temperature range where conduction is by variable-range hopping is not well understood. Evidence from the early work by Fritzsche is discussed by Shklovksii and Efros (1984), who come to the conclusion that the Hall mobility must be small. Hopkins et al. (1989) have investigated the behaviour of heavily doped Ge:Sb, pushed into the non-metallic regime by magnetic fields up to 7 T, at temperatures down to 100 mK. Below 1 K the

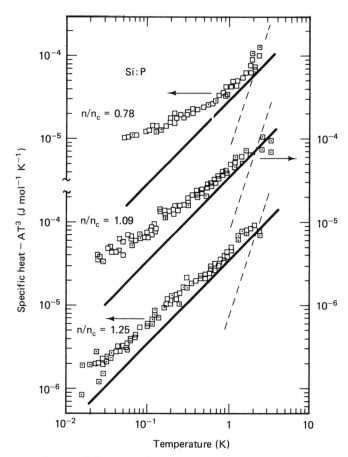

Fig. 5.11 Electronic part of the specific heat of Si : P as a function of temperature for three different doping levels n/n_c. The dashed lines represent the subtracted phonon contribution AT^3 ($\theta_D = 640$ K) and the solid lines are the expected specific heat $\gamma_0 T$ for degenerate electrons with effective mass $m^* = 0.34\,m$.

magnetoresistance increases by as much as 1000 as the field increases above the transition value, with the resistance varying as $\exp(bH^2/T^{1/4})$, with b constant. On the other hand, the measured Hall coefficient changes by only about 4. Other evidence suggesting this behaviour was quoted by Adkins (1978), who presented a speculative theory.

10 Specific heat near the transition

In Fig. 5.11 we show the electronic part of the specific heat of uncompensated Si : P for various values of $n - n_0$ (Paalanen *et al.* 1988). The full lines show the calculated specific heat for a degenerate gas of electrons with effective mass

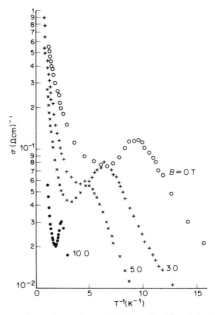

Fig. 5.12 Conductivity σ plotted against $1/T$ for Si:Sb with dopant concentration $2.7 \times 10^{18}\,\mathrm{cm}^{-3}$ for varying magnetic field B (Long and Pepper 1984).

$m_{\mathrm{eff}} = 0.34m$. Since in this range of concentration we believe that the Fermi energy lies in an impurity band, the observation that c_v tends to the free-electron value for T above 1 K is surprising; Kaveh and Liebert (1988) have shown, however, that this is to be expected. The anomaly at very low temperatures, showing an increase in γ by two or three has not been fully explained. Paalanen *et al.* (1988) account for this in terms of localized moments, which are in accord with the much greater increase in the magnetic susceptibility. We comment that free spins cannot exist in a metal (see Chapter 3, Section 8 on the Kondo effect). Another possible origin of the enhancement is the *increase* in the thermodynamic density of states resulting from the Altshuler–Aronov correction const $\times |E_c - E|^{1/2}$, which is positive for Si:P.

11 Effect of spin–orbit interaction

The effect on quantum interference of collisions in which the spin of the scattered particle changes sign has been described in Chapter 1, Section 8. Kaveh and Mott (1987) have discussed the effect of such scattering on the conductivity of doped silicon near the transition, where the conduction is in an impurity band produced by the dopant. If Z' is the atomic number of the matrix and Z that of the donor, spin–orbit interaction varies as $(Z - Z')^4$, and its effects will therefore be strongest for large Z. Kaveh and Mott suggest that the effect observed by Long and Pepper (1984) in Si:Sb should be attributed to spin–orbit scattering, and its absence in

Si:P or Si:As to the Z dependence. Their results are illustrated in Fig. 5.12. Kaveh and Mott introduce the spin–orbit scattering time τ_{so} and a new length scale $L_{so} = (D\tau_{so})^{1/2}$, and the Kawabata equation (9) becomes

$$\sigma = g^2\sigma_B\left[1 - \frac{C}{(gk_Fl)^2}\left(1 - \frac{l}{L_2}\right) + \frac{C_{so}}{(gk_Fl)^2}\frac{l}{L_{so}}\left(1 - \frac{L_{so}}{L_i}\right)\right], \quad L_i > L_{so},$$

where C_{so} is positive. The presence of this term will therefore push the mobility edge to lower energies, as long as $L_i > L_{so}$. As the temperature rises, L_i drops, and when $L_i = L_{ao}$ the mobility edge will rise to the value without spin–orbit scattering. Thus, as in Fig. 5.7 (measurements being made for $n < n_c$), at low temperatures the effective value of $E_c - E_F$ is smaller, giving higher conductivity than at higher temperatures, when spin–orbit effects are absent. This is what is observed in Fig. 5.12.

The effects of a magnetic field are discussed by Kaveh and Mott (1987).

12   Evidence that the transition lies in an impurity band, and that the two Hubbard bands have merged

Alexander and Holcomb (1968) first gave evidence to show that the metal–insulator transition in doped silicon or germanium occurs in an impurity band, which merges with the conduction band when $n \approx 3n_c$; the evidence comes from the changing nature of the Knight shift, being determined by the phosphorus atoms in the impurity-band regime. Further evidence along the same lines was given by Jérome et al. (1985). Much earlier, Mott and Twose (1961) supposed that the transition occurs in an impurity band. We have seen in Section 11 the curious coincidence that the effective mass in the impurity band in Si:P appears to be about the same as for the conduction band. This could hardly be so for compensated samples; measurements of γ for these are desirable.

The assumption that the transition takes place in an impurity band does not necessarily mean that there is a gap between it and the conduction band. It means that the wave functions ψ are such that, at the Fermi energy, $|\psi|^2$ is much greater in the dopant atoms than elsewhere. No sharp transition between the two situations is envisaged.

A further argument against a transition in the conduction band is given by the present author (Mott 1983), who maintains that in the conduction band it is impossible that the random situation of the donors should give disorder strong enough to produce Anderson localization.

In n-type GaAs Ming-Way Lee et al. (1988) used far-infrared optical absorption to show that, in the metallic state near the magnetic-field-induced transition, the 1s–2p absorption by donors persists, giving evidence that the transition lies in an impurity band.

The second question that we discuss in this section is whether the Hubbard U determines the concentration at which the transition takes place—a possibility that could occur only in an impurity band—or whether the transition is purely of

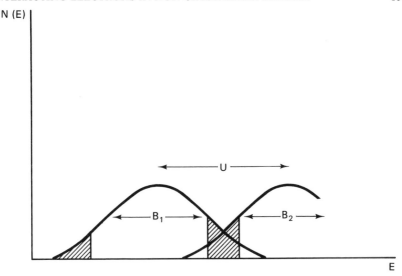

Fig. 5.13 Schematic illustration of how two Hubbard bands, with localized tails (shaded), resulting from disorder, can overlap, so that the equation $\frac{1}{2}(B_1 + B_2) \approx U$ determines approximately the concentration at which the transition occurs, while the properties of the materials near the transition are those resulting from a transition of Anderson type.

Anderson type (though affected of course by long-range interaction). Until recently it was supposed by the present author that the former is the case. We must now favour, however, the latter assumption for many-valley materials (e.g. Si and Ge), the Hubbard gap opening up only for a value of the concentration n below n_c. The first piece of evidence comes from a calculation of Bhatt and Rice (1981), who found that for many-valley materials this must be so. The second comes from the observations of Hirsch and Holcomb (1987) that compensation in Si : P leads to localization for a smaller value of n_c than in its absence. As pointed out by Mott (1988), a Mott transition occurs when $B = U$ (B is the bandwidth, U the Hubbard intra-atomic interaction), while an Anderson transition should be found when $B \sim 2V$, where V is some disorder parameter. Since $U \sim e^2/\kappa a_H$, where a_H is the hydrogen radius, and $V \sim e^2/\kappa a$, and since at the transition $a \approx 4a_H$, if the transition were of "Mott" type then it should be the other way round.

Zabrodskii and Zinov'eva (1984) found that in n-type germanium the transition can be induced by increasing the compensation, which is consistent with the above argument. They found hopping conduction with $v = \frac{1}{2}$ on the insulating side, with the Coulomb gap tending to zero at the transition, as noted in Chapter 1, Section 15

In earlier work (see e.g. Mott 1987) the present author has attempted to combine the hypothesis that the Hubbard U determines the value of n_c in doped semiconductors with the observation that the transition shows the properties of one of Anderson type (second order, $c_v \neq 0$, quantum interference and interaction effects) by supposing that two Hubbard bands, separated by U, have small localized tails, as in Fig. 5.13, and that the transition occurs for a value of n_c such that $U \sim B$, when states in the overlapping tails become delocalized at E_F. This is certainly not correct for Si and Ge, but may possibly be for single-valley materials. It is correct, however, for liquid Na_xNH_3, where the miscibility gap indicates a Mott transition, while $\sigma = \sigma_{min}$ at the concentration for which this occurs; this phenomenon is discussed further in Chapter 10, Section 5.

13 A degenerate gas of small polarons

In Chapter 2, Section 2 we have indicated that interaction with phonons could, in certain circumstances, give a very large mass enhancement throughout the conduction band of a metal, and in this case the electrons should be described as a degenerate gas of small polarons. Materials of this kind are $SrTiO_3$ and $KTaO_3$; these are wide-gap materials that, if stoichiometric, would be insulators, but that through oxygen deficiency can have a small concentration of free electrons, which form a degenerate gas in the d-band of Ti or Ta. $SrTiO_3$ was investigated by Frederikse *et al.* (1964); the conductivity tends to a finite value at low temperatures if the concentration of carriers exceeds 3×10^{18} cm^{-3}. The equation

$$n^{1/3}a_H \approx 0.2$$

would then indicate that the hydrogen radius is 1.5×10^{-7} cm. The static dielectric constant is large, $\kappa \approx 220$. Conductivity, thermopower and Hall effect indicate that $m_{eff} \approx 0.1m$; the hydrogen radius given by this value, $\kappa\hbar^2/m_{eff}e^2$, is 12 Å, about the same as that quoted above. Thus the low value of n to which metallic conductivity persists, in spite of the *high* effective mass, is explicable since the force between the electrons and the donor is $e^2/\kappa r^2$ down to distances much smaller than the hydrogen radius. That this must be so is also indicated by the very high mobility at low temperature, giving weak Coulomb scattering by the positive charges and a mean free path of order 800 Å. But this could only occur if the electron forms a small polaron, of radius much smaller than a_H; in other words, if the medium is fully polarized around the moving electrons as well as around the positive charge. For low concentrations a hydrogen-like pair must form between the positive charge and the polaron.†

Wemple (1965) and Wemple *et al.* (1966) have made similar observations for $KTaO_3$ in a range of donor concentrations from 3×10^{17} to 10^{19} cm^{-3}. The static dielectric constant is about 4500 at low temperatures and the mobilities are still higher (23 000 cm^2 V^{-1} s^{-1} at 4 K for the lowest concentration).

We conclude from these results that a degenerate gas of small polarons is a valid concept.

---

† There are several papers on the energy-band structure of $SrTiO_3$, e.g. Soules *et al.* (1972). Wolfram (1972) emphasized the two-dimensional character of the Ti d-band (the conduction band), giving a rapid rise of $N(E)$ with E. This will facilitate polaron formation.

14 Anderson versus band-crossing or Mott transitions

In this chapter we have seen how, by alloying an insulating material like silicon with small quantities of another material of different valence (e.g. phosphorus or boron), an impurity band can be formed and a metal–insulator transition occur as the concentration is increased. In the presence of compensation this is certainly of Anderson type, and, as we have seen, is probably so without compensation. If so then both the activation energy ε_2 for conduction and the zero-temperature conductivity itself go continuously to zero as the concentration is decreased. But small additions of a second material may produce a discontinuous Mott or band-crossing transition if the host is in a condition near to metallic behaviour; if so then σ should show a discontinuous drop to zero with increasing concentration of the second material. We do not know of a case where this has been described. As will be shown in Chapter 6, Ti_2O_3 seems to be a material where the admixture of V_2O_3 leads to a discontinuous band-crossing transition; and when V_2O_3 is alloyed with Ti_2O_3 there appears to be a discontinuous Mott transition. Also, the addition of Ti_2O_3 to V_2O_3 leads to a sharp drop in the activation for conduction.

Which sort of transition is occurring in the new superconductors, such as the antiferromagnetic insulator $La_{2-x}Sr_xCuO_4$ for small x, will be discussed in Chapter 9.

6

Metal–Insulator Transitions in Transitional-Metal Oxides

1 Introduction

In this chapter we describe a variety of transitions. Ti_2O_3 may possibly show a band-crossing transition when alloyed with V_2O_3, with the predicted discontinuity in the conductivity at about 2.6% of vanadium (Section 3). VO_2 is a classic example of an oxide with one-electron centres (V^{4+}), which forms pairs at low temperatures,† but in which the pairing disappears at 340 K, leading to an increase in σ by about 10^5. According to Zylbersztejn and Mott (1975), the insulating phase, though associated with the pairing, is not to be explained simply by the formation of covalent bonds, but involves the Hubbard U (Section 5). We also describe Verwey-type transitions as in Fe_3O_4. Then V_2O_3 and NiS_2 are, we believe, classic transitions of Mott–Hubbard type. The transition that occurs in the high-temperature superconductors such as $La_{2-x}Sr_xCuO_4$ will be discussed in Chapter 9, and some further transitions of indeterminate type such as in $La_{1-x}Sr_xVO_3$ will be considered in Chapter 7.

First, however, in view of its importance in the development of our understanding of these materials, we give a description of nickel oxide, together with MnO and similar materials.

2 Nickel oxide, cobalt oxide and manganese oxide

Before the discovery of antiferromagnetism, it was pointed out that, according to the Wilson theory of metals and insulators (Chapter 1), nickel oxide should be a metal—whereas it is a transparent insulator. The nickel ions should have eight d-electrons, and the only splitting of the d-band to be expected is into the e_g and t_{2g} bands, with four and six electrons. Peierls explained in 1938 how this

---

† Compare the properties of fluids such as Na_xNH_3 or Cs near the critical point, where this happens (Chapter 10).

might be due to electron–electron interaction (Mott 1980b). We described in Chapter 4, Section 7 how Slater (1951) showed that the antiferromagnetic sublattice in NiO could split the e_g band into two bands with two electrons per atom in each, and thus explain its insulating properties, but pointed out that there is no marked increase in the conductivity when the temperature is raised above the Néel temperature. One way of understanding this is to consider, in the sense of Hartree–Fock, that each electron moves in a field in which a strong repulsive potential operates on half the nickel atoms, selected at random; this would give rise to strong Anderson localization. The work of Brandow (1985, 1987) investigates this problem in some detail.

We discuss here, following Honig (1984), the conductivity properties of NiO, CoO and MnO. For MnO Pai and Honig (1981) made a serious attempt to obtain single crystals of high purity and perfection. For the conductivity they found $E_\sigma = 0.74$ eV and for the thermopower $E_S = 0.30$ eV. This would imply that the carriers form polarons with the very high hopping energy $W_H = 0.44$ eV. Crevecoeur and de Witt (1968) also found $E_S < E_\sigma$. For CoO (Joshi $et\ al.$ 1980) $E_\sigma = 0.74$ eV and $E_S = 0.306$ eV; again there is a large difference, suggesting polaron formation. Such a large value of $E_\sigma - E_S$ is also found in vanadium glasses (Mott and Davis 1979, p. 142).

It is conjectured that in those materials in which the t_{2g} band (separated from the e_g band by crystal-field splitting) is not full, the carriers lie in this band; the gap is that needed to form from two Mn^{2+}, a Mn^{3+} and a Mn^+ ion, for instance.

In NiO the position is less clear; the optical band gap $h\nu_0$ is about 4.3 eV (Terakura $et\ al.$ 1984), and conductivity measurements on pure NiO are hard, because the resistivity is high and measurements are very sensitive to surface conduction. Wittenauer and Van Zandt (1982) claimed $E_\sigma \approx \frac{1}{2}h\nu_0$, which would imply that polarons are not formed, though their finding that $E_S < E_\sigma$ suggests the opposite.

Whether the carriers remain within the e_g band is uncertain. Honig considers that excitation from the 3d- to the 4s-band may be a possibility, the upper Hubbard band for the e_g states lying at a higher energy. In this case no polarons would be expected for the electrons, but would for the holes. Thus we should expect $E_S = E_\sigma$, but both smaller than $h\nu_0$.

Much experimental work has been carried out on NiO doped with Li_2O. A lithium ion Li^+ replaces Ni^{2+} in the lattice, a neighbouring Ni ion having the charge Ni^{3+}. A bound hole is then created. Bosman and Crevecoeur (1966) found that E_S and E_σ were identical (Fig. 6.1a), suggesting that polarons are not formed.

At low temperatures a drop in the thermopower occurs (Fig. 6.1a), suggesting that impurity conduction is present. This is also confirmed by a very marked drop in the activation energy in the conductivity (Fig. 6.1b). A very interesting phenomenon is observed in this material. Kolber and MacCrone (1972) found that this absorption of energy by $bound$ polaron hopping, from the motion of a hole round the lithium ion or Ni^{2+} vacancy, is 0.7 eV, suggesting a very large polaron hopping energy. The effective bandwidth for such hopping to nearest neighbours of the defect will of course be less than for free motion, giving stronger

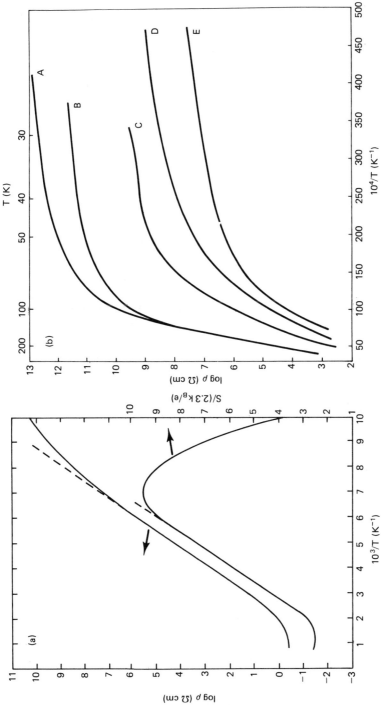

Fig. 6.1 (a) Logarithm of resistivity and thermopower $S/(2.3k_B/e)$ of Ni doped with 0.088% Li_2O. (b) Conduction in crystalline NiO; logarithm of resistivity as a function of $1/T$. The values of x in the formula $Li_xNi_{1-x}O$ were as follows: A, 0.002; B, 0.003; C, 0.018; D, 0.026; E, 0.032.

polaron formation. According to Kuiper *et al.* (1989), the hole lies in the oxygen 2p-band. However, properties of lithium-doped NiO are difficult to interpret. According to Honig, the properties of the host, particularly in polycrystalline specimens, are greatly affected by the lithium, and d-bandwidths are perhaps widened.

It is of course possible that a carrier in the conduction band or a hole in the valence band will form a spin polaron, giving considerable mass enhancement. The arguments of Chapter 3, Section 4 suggest that the effective mass of a spin polaron will depend little on whether the spins are ordered or disordered (as they are above the Néel temperature T_N). This may perhaps be a clue to why the gap is little affected when T passes through T_N. If the gap is $U - \frac{1}{2}(B_1 + B_2)$, and B_1 and B_2 are small because of polaron formation and little affected by spin disorder, the insensitivity of the gap to spin disorder is to be expected.

On the other hand, hopping conduction in antiferromagnetic insulators seems to be greatly affected by spin disorder, as shown by the work of Whall *et al.* (1987) on $MnFe_2O_4$, in which both $d\sigma/dT$ and dS/dT show striking changes at the Néel temperatures (about 600 K).

For nickel oxide, the fact that band-structure calculations have not been able to predict a band gap greater than about 0.2 eV (Sawatzky and Allen 1984, Huefner 1985) shows the importance of the Hubbard U. According to the work of the Groningen school, based on experimental work on X-ray absorption and photoemission, the upper Hubbard band is narrow and d-like, the lower band heavily hybridized with oxygen 2p, which has the result that in, for instance, NiS as compared with NiO, the Hubbard U is strongly dependent on the anion (Van der Laan *et al.* 1986, Zaanan and Sawatzky 1987). These authors describe such materials, in contrast with "Mott–Hubbard" insulators such as V_2O_3, as "charge-transfer insulators".

In NiO a metal–insulator transition has been observed under very high pressure (2.5 Mbar) (Kawai and Mochizuki 1971); the conductivity at room temperature dropped abruptly by about 10^6. Nothing was determined regarding the temperature dependence of the conductivity, or the change of volume.

3   Titanium trioxide (Ti_2O_3)

This material, which has the corundum structure, is a semiconductor at low temperatures, the optical band gap being 0.2 eV (Lucovsky *et al.* 1979). We should probably consider it to be an intrinsic semiconductor, but the activation energy in the conductivity does not appear to be constant. The thermopower (Chandrasekhar *et al.* 1970) is about $900 \, \mu V \, K^{-1}$ at 100 K and $500 \, \mu V \, K^{-1}$ at 200 K; this would suggest an activation energy of about 0.06 eV, or less than half the band gap.† This makes it likely that one of the carriers is a small polaron; the

---

† Shin *et al.* (1973) give a value (half the gap) between 0.05 and 0.1 eV.

thermopower is p-type, so it must be the electrons that form them; the current is carried mainly by the more mobile positive holes.

The resistivity shows, with increasing temperature, a *gradual* transition between about 400 and 500 K to a metallic behaviour with conductivity in the metallic range $10^3–10^4 \, \Omega^{-1} \, cm^{-1}$. In this range there is a change in the ratio c/a, together with the disappearance of the optical gap. An early theory (see e.g. Mott 1969) was that the gap was a function of c/a, and that it would shrink because, by allowing an increasing number of free electrons and holes, it would by so doing increase the entropy. Such a model is similar to that of Falicov and Kimball (1969) for other transitions that occur with increasing temperature.

However this model has had to be abandoned in view of the work of Dumas (1978) and Dumas and Schlenker (1976, 1979a, b) on the properties of the alloys $Ti_{2(1-x)}V_{2x}O_3$. The addition of V_2O_3 increases the conductivity, gives rise to variable-range hopping and eventually (for $x=0.1$) leads to a transition to the metallic state. c/a does change, reaching the value observed for pure Ti_2O_3 above 400 K when $x=0.1$; however, the increase in conductivity at about 400 K is still observed. The additional conductivity is caused by an impurity band. The vanadium (at any rate for small x) is in the state $3d^3$ (positively charged), and, to preserve charge neutrality, defects are formed at which a hole is localized, giving Ti^{4+} ions. An impurity band of this kind probably exists in many undoped Ti_2O_3 crystals, on account of lack of stoichiometry.

The gap, then, does not alter with temperature. Mott (1981) suggested that, on account of the large effective mass of the electrons, calculated to be $m_{eff} = 6m$ even without polaron formation (Ashkenazi and Chuchens 1975), a transition will occur when $-F$, where F is the *free* energy of the electron–hole gas, is equal to ΔE per pair, ΔE being the band gap. $|F|$ will increase with T. In order to understand the range ΔT of $T (\sim 100 \, K)$ over which the transition takes place, one could perhaps assume that point charges resulting from a lack of stoichiometry give rise to fluctuations in the positions of the bottom of the conduction band. One needs a fluctuation $k_B T$ when $T \sim 100 \, K$, and thus $\sim 0.007 \, eV$. If point charges are at a distance R apart, this should be equal to $e^2/\kappa R$. For the dielectric constant $\kappa \approx 45$ (Lucovsky *et al.* 1979), so $R = 50 \, Å$, giving a concentration of point charges of order $10^{19} \, cm^{-3}$, which seems reasonable.

On the other hand, such a model would suggest an increase in ΔT for a specimen alloyed with V_2O_3, whereas the reverse seems to be the case (see Mott 1981). This may be due to an increase in the dielectric constant resulting from the formation of an impurity band, which decreases the term $e^2/\kappa R$.

Chen and Sladek (1978) investigated the piezoresistance of Ti_2O_3, deduced deformation potentials for the holes and came to the conclusion that the overlap between the valence and conduction bands is never large.

It would be of interest to have measurements at lower temperatures to see if $\sigma(T=0)$ goes continuously to zero with increasing vanadium, or whether there is a discontinuity. A concentration 0.25% of V does not change the slope of the $\ln \sigma - 1/T$ plot appreciably; it seems possible that V does not decrease the gap until a spontaneous metal–insulator transition of band-crossing type occurs, the conductivity jumping to a value determined by the (small) number of electrons and by scattering of the electrons by the V atoms.

4 Vanadium sesquioxide (V_2O_3)

4.1 Introduction

The well-known semiconductor-to-metal transition for this material, which occurs with increasing temperature, is shown in Fig. 6.2; this was first observed by Foëx (1946). The metallic phase has the corundum structure and is not ferro- or antiferromagnetic (Gossard et al. 1970, Wertheim et al. 1970). There is a small decrease in volume ($\sim 1\%$) on going to the metallic phase, so that, as first shown by Austin (1962), the transition temperature can be lowered by pressure, and above 24 kbar the metallic phase is stable down to the lowest temperatures (McWhan and Rice 1969). The low-temperature insulating phase is anti-ferromagnetic (AF) with a moment of about $1.2\mu_B$ (Moon 1970, Andres 1970); its crystal structure will be discussed later. The addition of Ti^{3+} or Mg^{2+} ions to V_2O_3 leads to a suppression of the AF insulating phase, whereas the addition of Cr^{3+}, Ti^{4+} and Fe^{3+} results in a first-order transition to the AF insulating state (McWhan and Remeika 1970, Menth and Remeika 1970, McWhan et al. 1973b). In the last reference it is suggested that the simplest explanation is that the Cr^{3+} ion is a localized impurity and that it deletes a state from the d (e^T and a^T) bands. Deleting a state, according to Brinkman and Rice (1973), is equivalent to a band-narrowing that drives the system towards the insulating state. Likewise, Ti^{3+} adds a slightly larger e^T orbital and therefore causes an increase in bandwidth. Figure 6.3 shows the phase diagram and hence the dependence of critical pressure on composition; it is not clear why the curve goes smoothly from Cr to Ti.

As Fig. 6.2 shows, the transition in this material and in its alloys can be subject to hysteresis. This has been cited by some authors (e.g. Honig 1985) as evidence that the transition is not a "Mott transition". As explained in Chapter 4, however, we define the Mott transition as one from an insulating state with lowest free energy and an antiferromagnetic ordering of moments to a metallic state without, and for the two states the crystal parameters may well be different. We emphasize that for the insulating state band-theory calculations without the anti-ferromagnetic ordering of moments should *not* give a gap at all. If the calculations give a gap (unless very small), antiferromagnetism should not occur.

4.2 The metallic phase of V_2O_3

Metallic V_2O_3 at the lowest temperatures can be stabilized by pressures of 23 kbar or by the addition of Ti_2O_3. The most striking property of the metallic phase at low temperatures is the high linear specific heat γT, with $\gamma = 96 \times 10^{-4}$ cal K^{-2} mol^{-1} (McWhan et al. 1971), and the high Pauli

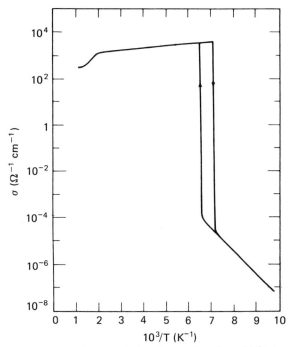

Fig. 6.2 Conductivity of a single crystal of V_2O_3 as a function of $1/T$, showing the metal–insulator transition (Föex 1946).

paramagnetism. These large specific heats can be obtained when the metallic phase is stabilized by alloying (McWhan *et al.* 1973a) or under pressure, where McWhan *et al.* 1973c) have made observations down to 0.3 K at 25 kbar.

Some specific heats are shown in Fig. 6.4. They are about three times that for palladium and would imply a bandwidth of 600 K ($\approx \frac{1}{20}$ eV) if interpreted by a model without correlation. Figure 6.5 (Mott and Jones 1936) shows that C_v should tend to a constant value at about one-third of the degeneracy temperature.

At higher temperatures the observed entropy change at the transition ($\Delta S = 2.6$ eu) at 1 atm is nearly equal to γT_c ($T_c = 150$ K). Although the value of S should be somewhat less than that given by $S = \gamma T$, because the specific heat is not necessarily linear, there can be no doubt that a major part of the entropy is electronic.

The magnetic susceptibility is also very high (Menth and Remeika 1970); the density of states deduced from the specific heat, and that deduced from the susceptibility, are in the ratio

$$N(E)_\chi / N(E)_\gamma = 1.8.$$

The magnetic behaviour of V_2O_3 and of the Ti-doped alloys is shown in Fig. 6.6. It can be seen that the susceptibility is higher in the metallic than the non-metallic phase. The drop in the susceptibility with temperature above about

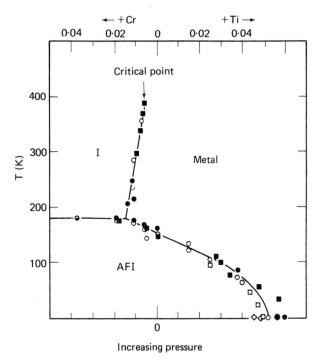

Fig. 6.3 Generalized phase diagram of the metal–insulator transition in V_2O_3 as a function of doping with Cr or Ti and as a function of pressure, showing the critical point (McWhan *et al.* 1971).

100 K suggests a very high effective mass, or a Fermi energy for the carrier of the order $k_B T$, where $T = 300$–600 K.

The electrical properties can be summarized as follows.

Austin and Turner (1969) found an n-type Hall coefficient at room temperature (300 K), the value corresponding to 0.6 carriers per vanadium atom (if they are all of one sign, though less if the behaviour is "semimetallic"). The conductivity of their specimens was $2.5 \times 10^3 \, \Omega^{-1} \, cm^{-1}$ and $\mu_H = 0.6 \, cm^2 \, V^{-1} \, s^{-1}$. The thermopower was p-type and $\sim 12 \, \mu V \, K^{-1}$. The discrepancy in sign suggests two kinds of carrier, as discussed by these authors. Somewhat similar results were obtained by Zhuze *et al.* (1969). McWhan *et al.* (1973b) reported on measurements on V_2O_3 at 20 kbar and liquid-helium temperatures, giving a slightly higher value of $R_H(3.5 \times 10^{-4} \, cm^3 \, C^{-1})$ in contrast with 2.4×10^{-4} at 1 atm and 150 K.

The T^2 term in the resistivity ρ of metallic V_2O_3 above 20 kbar, first observed by McWhan and Rice (1969), is shown in Fig. 6.7. It will be noticed that there is some tendency for ρ to saturate above 300 K as for Landau–Baber scattering with a low degeneracy temperature (Chapter 2).

The effect of impurities on the resistivity of metallic V_2O_3 is very large. Thus McWhan *et al.* (1973b) found 140 $\mu\Omega$ cm per atomic per cent of Cr for the residual

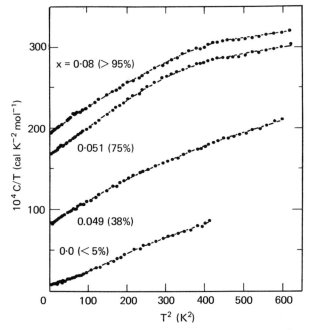

Fig. 6.4   Specific heat of $(V_{1-x}Ti_x)_2O_3$ divided by T plotted against T^2. The first number on each line shows x, the second the proportion of the compound in the alloy phase (McWhan *et al.* 1971).

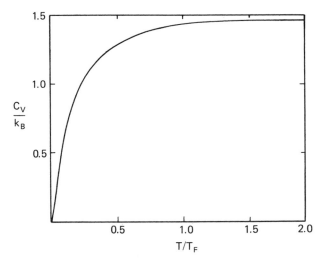

Fig. 6.5   Specific heat of a degenerate gas as a function of temperature when the degeneracy temperature is exceeded (Mott and Jones 1936).

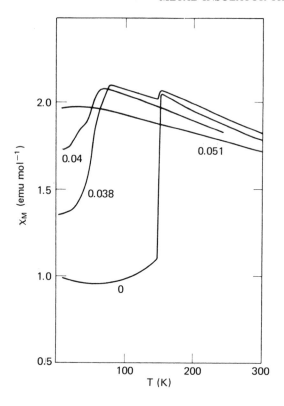

Fig. 6.6   Susceptibility of $(V_{1-x}Ti_x)_2O_3$ for various values of x shown on each curve, as a
function of T (Menth and Remeika 1970).

resistance of V_2O_3 under pressure at low T. These authors point out that this is
more than an order of magnitude greater than one expects from the formula

$$\sigma = S_F e^2 l / 12\pi^3 \hbar$$

if $l^{-1} \gtrsim NA$; here N is the number of chromium atoms per unit volume and A
the cross-section. In our view, this is to be explained by the considerations of
Chapter 4, Section 10.

Figure 6.8 shows the reflectivity of crystals of V_2O_3 obtained by Fan (1972).
From the low-frequency behaviour, Fan deduced $m \approx 9m_e$; since for a bandwidth
~ 1 eV, $m_{eff} \approx m_e$, the factor 9 represents the mass enhancement. The inflexion
above 1 eV may be due to the transition to the unenhanced mass discussed in
Chapter 4, Section 6. Results for VO_2 are also shown in Fig. 6.9; they give $m \approx 3m_e$.

We think that all these observations are to be explained by the assumption
that metallic V_2O_3 is a highly correlated electron gas, as first suggested by
Brinkman and Rice (1970b) and described in Chapter 4. The very low degeneracy
temperature suggests that there may also be some mass enhancement of the
carriers by polaron formation. Two electrons per atom would just half fill an e_g^T
band, so that the number of electron-like and hole-like carriers would be

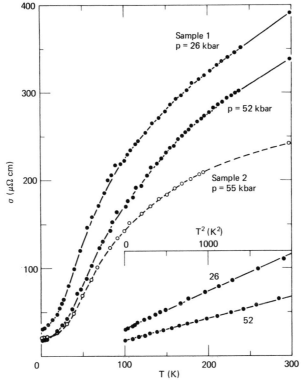

Fig. 6.7   Resistivity of metallic V_2O_3 under pressure as a function of temperature (McWhan and Rice 1969).

identical; the p-type thermopower observed must be due to overlapping of the two bands. Discussions of the band structure by Kawamoto *et al.* (1980), Honig (1985) and other earlier work cited in these papers support this.

It is thought that two bands are partly occupied, the a_{1g}^* with space for two electrons per V atom, and e_g with four, the former with lobes of d-functions along the c-axis and tending to decrease c/a, the latter with lobes in the basal plane tending to increase it. Figure 6.10 shows the behaviour of c/a for rhombohedral V_2O_3 as a function of temperature, and also the behaviour at the transition (above the Néel temperature for the insulating phase). We interpret this by supposing that the wave functions of the two bands are mixed in such a way as to minimize the free energy, and that the entropy is increased by increased mixing. The large c/a ratio for metallic V_2O_3 at the transition indicates, as Goodenough (1971) points out, that the a_{1g} band must be above the $e_g(\pi^*)$, in contrast with Ti_2O_3. As the temperature rises, it comes down, leading to a decrease in c/a.

In our view (Mott 1972b), mixing the bands increases the tendency to moment formation and to Mott–Hubbard insulation, because, while U is

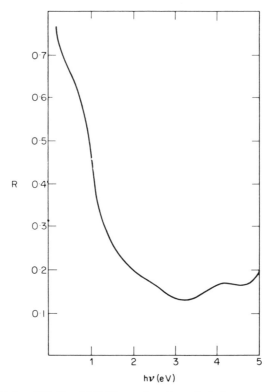

Fig. 6.8 Reflectivity of V_2O_3 at 300 K as a function of photon energy $h\nu$ (Fan 1972).

unchanged, the cost in kinetic energy of localizing the electrons is decreased. Since in the metallic phase the nearer we are to the transition the greater is the entropy, the degree of mixing should always increase with increasing temperature. The change of c/a, then, is a consequence of the increased hybridization of the wave functions.

A much smaller value of c/a for $(V_{0.96}Cr_{0.04})_2O_3$ is observed and suggests that at the temperature of the transition the bands are already well mixed, and not much entropy is to be gained by mixing them further. This may be one reason, quite distinct from band-narrowing, why Cr stabilizes the insulating phase.

4.3   The insulating phase of V_2O_3

The structure of the insulating low-temperature phase is a monoclinic distortion of the corundum structure. Structure determination (Dernier and Marezio 1970) shows that the V–V distance increases from 2.697 to 2.745 Å, which across shared octahedral edges increases from 2.882 to 2.907. The average V–O distance remains constant. The material is antiferromagnetic, with a moment of $1.2\mu_B$ on

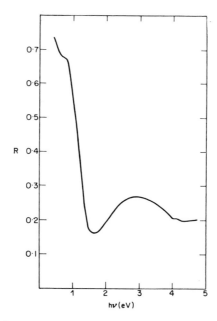

Fig. 6.9   Reflectivity of VO_2 at 360 K (Fan 1972).

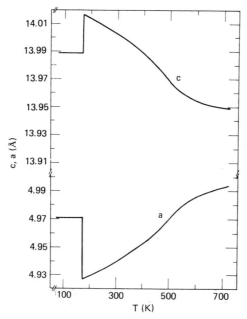

Fig. 6.10   Lattice parameters c and a for V_2O_3 with corundum structure (McWhan and Remeika 1970).

each atom. The interpretation has been widely discussed; we refer the reader to Goodenough (1971, p. 291). We think that the system is a normal antiferromagnetic non-metal (like NiO) and that the distortion may be a form of magnetostriction. It disappears above the Néel temperature, which is somewhat below 200 K (cf. Fig. 6.3), the structure then being corundum. In this phase we consider the material as simply a magnetic insulator above its Néel temperature, the Hubbard gap remaining as usual unchanged.

4.4   The phase diagram and the metal–insulator transition

The observation that the addition of Cr is equivalent to a negative pressure has already been mentioned. The phase diagram of $(V_{1-x}Cr_x)_2O_3$ as a function of x or pressure due to McWhan et al. (1971) was shown in Fig. 6.3. The antiferromagnetic insulating phase is monoclinic, as already stated. Ashkenazi and Weger (1973) accounted for the metal–insulator transition from the metallic to this monoclinic phase solely in terms of an (unrestricted) Hartree–Fock model, which, according to the considerations of Chapter 4, Section 6, must be acceptable at zero temperature. In this book, however, we shall not discuss the monoclinic phase, which is clearly bound up with long-range antiferromagnetic order. In going above the "Néel temperature" to the paramagnetic insulating state, we go over to a corundum structure; this seems to us to show that the insulating phase is not bound up with the monoclinic distortion, but is a consequence of the long-range magnetic order—or above the Néel temperature it is connected with the existence of moments. A metal–non-metal transition can occur in the corundum phase alone, being associated with a decrease of c/a in the insulating phase. This we have supposed is because decreasing c/a mixes the two bands, making Mott–Hubbard localization less costly in energy. There is only a small entropy change in crossing the boundary from the paramagnetic insulator to the metal, presumably because the moments are nearly random in both phases at these temperatures. In the insulating phase, because of the smaller number of current carriers, the entropy however should approach nearer to $k_B \ln 3$ per moment, which accounts for the slope of the line separating the two phases.

The critical point at 400 K is discussed by Jayaraman et al. (1970a). There should of course be a two-phase region if there is a critical point; this is not shown in Fig. 6.3. A two-phase region is a consequence of the "kink" in the free-energy curve shown in Fig. 4.2. The position of the critical point appears controversial; Kerlin et al. (1973) have produced evidence that the two-phase regions reach above 400 K, the transformation time from one phase to the other being very slow in pure V_2O_3.

Figure 6.11 shows the resistivities of some alloys $(V_{1-x}Cr_x)_2O_3$ as functions of temperature. The addition of Cr, as already stated, increases the temperature at which the transition to the metallic state occurs. At higher temperatures a transition back to the semiconducting state is predicted by the phase diagram of Fig. 6.3. Particularly remarkable, however, are the very low conductivities

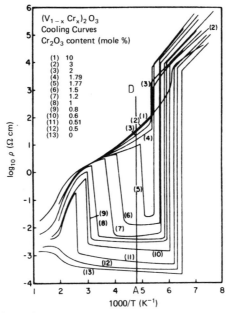

Fig. 6.11 Variation of resistivity with temperature in the $(V_{1-x}Cr_x)_2O_3$ system, $0 \leqslant x \leqslant 0.1$ (after Kawamoto et al. 1980). From Honig (1985).

reached at the transition to the metallic state for the higher values of x. This must mean that in this state the overlap between the two bands is very small indeed (see Chapter 1, Section 16), as pointed out by Honig.

A prediction of theory (Chapter 4) is that when an insulator–metal transition of either band-crossing or Mott type occurs through a change of composition in an alloy, the zero-temperature conductivity should jump discontinuously from zero to a finite value. This seems to be the case for the alloys with Ti_2O_3. The alloys with titanium have a conductivity when metallic of order $10^4 \, \Omega^{-1} \, cm^{-1}$ at low temperatures, for 5% titanium. Since the Ioffe–Regel value is about one-third of this, each titanium atom appears to have a scattering cross-section of about seven times its area. This is an example of the large scattering by impurities already referred to in Chapter 4, and characteristic, in our view, of the highly correlated electron gas.

5 Vanadium dioxide (VO_2)

5.1 Introduction

The transition in this material from a low-temperature semiconductor to a metallic phase at 340 K was first observed by Morin (1959); more recently, jumps in the conductivity by a factor as large as 10^5 have been observed in various laboratories (Paul 1970). The transition is first-order; the entropy change is $\Delta S = 1.6\,k_B$ per vanadium atom (Bergland and Guggenheim 1969). In the high-temperature phase VO_2 has a rutile tetragonal structure, each vanadium atom being located at the centre of an oxygen tetrahedron. The low-temperature phase is a monoclinic distortion of this rutile structure, involving both a pairing and an off-axis displacement of alternate vanadium atoms. The vanadium atoms, in the state V^{4+}, has one d-electron per atom; atoms with one outer electron form pairs in liquids (Chapter 10) or in solids if the structure allows it. Goodenough (1963) proposed a model in which this pairing is responsible for the insulating state, the gap being solely a consequence of the crystal structure.

This model was criticised by Zylbersztejn and Mott (1975), using information obtained from the properties of $V_{1-x}Cr_xO_2$. They considered the gap to be essentially of Hubbard type, and the pairing to affect it only slightly, though destroying the antiferromagnetism that would otherwise be present. The wave functions for the electrons in a pair could be of the London–Heitler type, namely

$$[\psi_a(1)\psi_b(2) \pm \psi_a(2)\psi_b(1)][\chi_\uparrow(1)\chi_\downarrow(2) \mp \chi_\uparrow(2)\chi_\downarrow(1)],$$

where ψ and χ are orbital and spin functions. When we say that this is a Hubbard gap, we mean that it is nearly the same as it would be if the material were antiferromagnetic, and arises because an electron in the conduction band is in the state V^{3+}, is mobile and is jumping from atom to atom, while a hole is also in a mobile state V^{5+}; the Hubbard U measures the energy difference between these states. The gap is about 0.8 eV, and the conduction band is of π^* type, about 0.1 eV wide. Conductivity and thermopower show the same activation energies, so no dielectric polaron is formed. This is surprising, because the high-frequency dielectric constant κ_∞ is estimated to be about 5, while $\kappa \approx 30$, so a significant polaron hopping energy is expected. It may be that this is present, but that in a material with a high Debye temperature we are in the range where polarons lead to mass enhancement rather than hopping.

The metallic phase is, according to this description, a result of the disappearance of the sideways displacement of the vanadium ions. The stronger overlap between the orbitals produces a metallic phase. The material is strongly magnetic, and has a large electronic specific heat ($\gamma = 3.4 \times 10^{-3}\,\mathrm{cal\,K^{-2}\,mol^{-1}}$, about 20 times that for silver. Some mass enhancement is probable. Since the electrons are in a state near a metal–insulator transition of Mott type (which does not occur because a structure change preempts it), this may be an example of the highly correlated gas of Brinkman and Rice (Chapter 4, Section 6). Zylbersztejn and Mott considered that the high effective mass is caused simply by a narrow band, but the evidence for this does not seem to be conclusive.

Deducing the electronic entropy change from the specific heat, it appears that about one-third is due to the electrons. The rest must be the result of soft phonon modes.

5.2 Effect of alloying

The addition of 2.5% CrO_2 leads at intermediate temperature to a phase in which only half the V ions are paired; the others form a zig-zag chain (Marezio et al. 1972, Pouget et al. 1974). At low temperatures pairing takes place, and at higher temperatures the usual transition to the metallic rutile form. This intermediate phase has high susceptibility, and the zig-zag chains are interpreted as one-dimensional Mott–Hubbard insulators above their Néel temperature. Since the transition temperature is little changed, this shows that U is the most important quantity in determining the gap.

The system $V_{1-x}Nb_xO_2$ has been investigated by Villeneuve et al. (1972), Pouget et al. (1972) and Lederer et al. (1972). These authors show that for values of x below 0.15 the low-temperature paired VO_2 structure persists; the Nb transfers its electron to a neighbouring V site, giving a V^{3+}–Nb^{5+} pair. This acts as a free spin giving Curie susceptibility at low temperatures. For $x > 0.15$, all spins are free at high T, giving approximate Curie behaviour. Although the structure is now rutile throughout (no pairing), the remains of a "transition" are observed, particularly in the thermopower, as shown in Fig. 6.12. It is particularly striking that the thermopower appears metallic when an activation energy remains for the conductivity as shown in Fig. 6.13.

Our interpretation is slightly different from that given by the authors. We think that there is some pairing of cations (V–V pairs), but the interruption of chains by the Nb atoms inhibits any structure change. Particularly interesting is the result at high temperatures (or large x, when the transition is hardly perceptible) that the susceptibility follows the Curie law. This means that the moments (V^{4+}, V^{3+}) are very weakly coupled to each other. In V_2O_3 stabilized by Cr_2O_3 in the AFI phase the $1/\chi$–T curve is far from the Curie straight line; the antiferromagnetic coupling must be weaker in the V–Nb oxide discussed here.

The coupling that in pure VO_2 prevents a large paramagnetic susceptibility is of course a consequence of this movement of the states V^{3+} and V^{5+}, present because rutile VO_2 is metallic. This cuts down the lifetime of the moments. In the mixed oxides with Nb both the Curie paramagnetism and the semiconductor-like behaviour show that this motion is strongly inhibited. Following Villeneuve et al., we think that this is due to Anderson localization resulting from the disorder (perhaps the one-dimensional nature of the motion along the chains promotes this). And once Anderson localization occurs, polaron formation must make the localization stronger. So the carriers are stationary, and the lifetime of the moments becomes long.

The heat capacities of various other vanadium oxides have been investigated by McWhan et al. (1973a). V_7O_{13} is interesting in that it is antiferromagnetic and remains metallic down to liquid-helium temperatures. γ is high, about half that for metallic V_2O_3, confirming our conjecture (Chapter 4, Section 6) that mass enhancement should exist for such materials. $V_{0.86}W_{0.14}O_2$ also remains metallic down to liquid-helium temperatures, retains the rutile structure and has a large value of γ (about two-thirds that for V_2O_3). We conjecture that this might

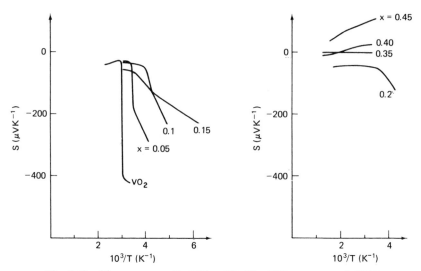

Fig. 6.12 Thermopower S of $(V_{1-x}Nb_x)O_2$ (Villeneuve *et al.* 1972).

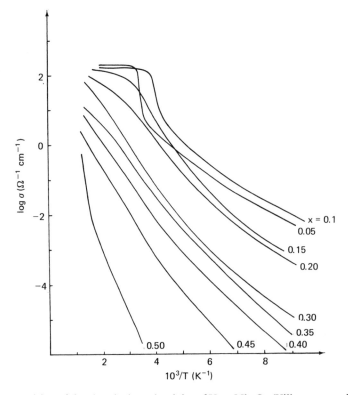

Fig. 6.13 Logarithm of the electrical conductivity of $V_{1-x}Nb_xO_2$ (Villeneuve *et al.* 1972).

be explained in the following way. The compound could be an antiferromagnetic insulator were it not for the extra electron on the W^{4+} ion compared with V^{4+}. These electrons lower the energy of the metallic phase, leading eventually at zero temperature to a metal–insulator transition as χ increases. Near to this the electron gas should be highly correlated and γ large.

6   Nickel sulphide (NiS)

This material, which has the nickel arsenide structure, shows a first-order transition as T increases, with a jump in the conductivity by a factor of about 40 at 260 K. The low-temperature phase is antiferromagnetic with moment about $1.7\mu_B$; the high-temperature phase shows Pauli paramagnetism. There is a decrease in volume at the transition, but, as far as is known, no change in structure. The transition temperature drops with pressure as shown in Fig. 6.14 according to McWhan et al. (1972); the high-temperature phase can be stabilized at pressures above 20 kbar, and the resistivity shows a normal temperature dependence over the whole range. This is interpreted as showing that there is no ferro- or antiferromagnetic ordering.

It is well established that the temperature, 260 K, at which the moments disappear is *not* a Néel temperature; magnetic and Mössbauer studies locate the Néel temperature much higher, above 1000 K (White and Mott 1971, Lowde et al. 1973) from the T-dependence of the susceptibility. So the antiferromagnetic coupling term I^2/U is likely to be larger than that in NiO; since I is likely to be larger but U comparable, this is not surprising. The large overlap between the orbitals with antiparallel spins could account for the value $1.7\mu_B$ lying below that for the Ni^{2+} ion $(2\mu_B)$.

The change in the susceptibility at the transition is quite small, as shown in Fig. 6.15.

As regards the electrical properties of the low-temperature phase, NiS, unless stoichiometric, remains conducting down to liquid-helium temperatures with a highly anisotropic conductivity $\sim 10^2\,\Omega^{-1}\,cm^{-1}$ independent of temperature. A "metallic" behaviour is also shown by the thermopower S, with S increasing with T, the results of Townsend et al. (1971) being reproduced in Fig. 6.16. It has now been shown, however, that for the stoichiometric material a gap appears between the two Hubbard bands (Ovchinnikov 1982). For specimens deficient in nickel it is a p-type semiconductor, a metal–insulator transition occurring for a low concentration. It has been suggested that in either case the carriers form spin polarons (Mott 1974a, Licciardello et al. 1977). White and Mott (1971), quoting the results of McWhan et al. shown in Fig. 6.14, argue for the latter situation.

The low-temperature phase under *pressure* shows a fairly high conductivity, as one would expect from increasing overlap of two Hubbard bands.

If, then, the low-temperature phase is semimetallic, we cannot interpret the first-order change at 260 K as a "Mott transition" caused by a discontinuous change in the number n of current carriers; that theory predicts a discontinuous

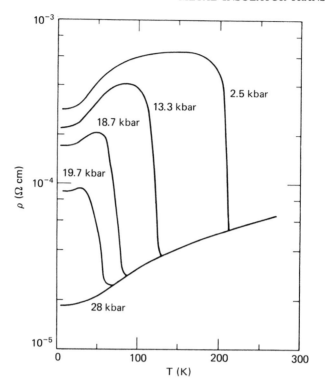

Fig. 6.14 Resistivity ρ of NiS as a function of T for various pressures (McWhan *et al* 1972).

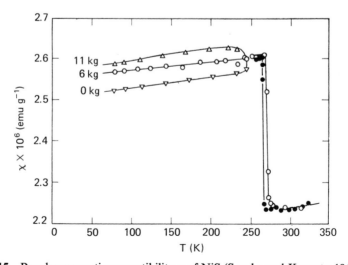

Fig. 6.15 Powder magnetic susceptibility χ of NiS (Sparks and Komoto 1968).

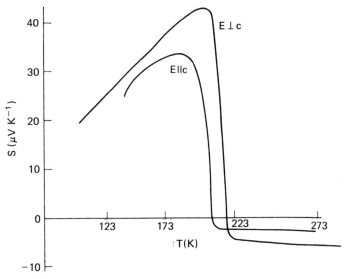

Fig. 6.16 Thermopower of NiS as a function of T (Townsend *et al.* 1971).

change in n from zero to some finite value, but not from one finite value to another. We think that the correct way to treat NiS is to postulate a discontinuous change in the moment (from $1.7\mu_B$ to zero), and treat the change of conductivity as a consequence of this. We have seen in Chapter 3 (Fig. 3.10) that the curve plotting energy against moment could have two minima, and we think that this is what happens in NiS; another example $(Co(Se_{1-x}S_x)_2)$ is discussed in Section 7.2. How this could happen can be seen in the light of a band structure first proposed by White and Mott (1971) and in principle confirmed by the calculation of Kasowski (1973). The band structure according to the former authors is such that a 3p-band from the sulphuric ions S^{2-} overlaps the conduction band formed from nickel in the state $3d^8(e_g)$. The result of Trahan (1972) (Fig. 6.17) showing an absorption edge at about 0.4 eV is in our view due to a transition from this 3p-band to the Fermi limit.† The existence of moments will depend on how strongly the 3p (sulphur) orbitals are hybridized with the 3d (nickel) orbitals. If we write these hybridized functions as

$$\psi_{3d} + M\psi_{3p}$$

then it is quite probable that two minima in the energy exist as a function of M; for one, for which M is large, U is correspondingly small because of the increased radius of the orbital, and the material is a normal metal; for the other, M is small, U is large and moments are formed on the nickel ions.

---

† Compare the absorption edge of metallic copper or silver due to transitions from the full d-band to the Fermi limit.

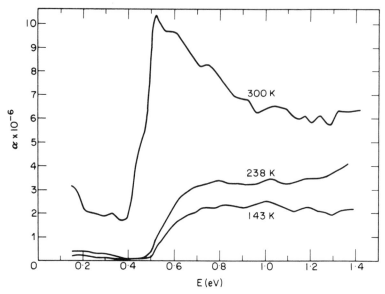

Fig. 6.17 Infrared absorption spectrum absorption coefficient of NiS. From Trahan (1972).

In comparison, FeS and CoS are antiferromagnetic metals at all temperatures. The substitution of Se or especially Te for S, by increasing the hybridization, stabilizes the non-magnetic metallic form.

Trahan and Goodrich (1972) found an entropy change at the transition in NiS of $5.03 \, \mathrm{J \, mol^{-1} \, K^{-1}}$. On the other hand, they found little change in the specific heat at the transition (as in VO_2), estimating that it increases by only $0.61 \, \mathrm{J \, mol^{-1} \, K^{-1}}$. If this is correct then the entropy in the metallic phase must be due to phonons, as in VO_2. This is confirmed by the work of Coey *et al.* (1974) and Brusetti *et al.* (1980). The latter authors point out, however, that the softening of c_{44} is to be ascribed to the change of structure. Coey *et al.* found a linear specific heat in the low-temperature phase with $\gamma = 2 \times 10^{-4} \, \mathrm{cal \, mol^{-1} \, K^{-2}}$ $(0.09 \, \mathrm{mJ \, mol^{-1} \, K^{-2}})$ and in the metallic phase stabilized by non-stoichiometry with $\gamma = 17 \times 10^{-4} \, \mathrm{cal \, mol^{-1} \, K^{-2}}$. This confirms the semimetallic nature of the low-temperature phase. No more than 25% of the entropy of the transition is electronic, the rest being due to phonons. The susceptibility of the metallic phase is about $2 \times 10^{-4} \, \mathrm{emu \, mol^{-1}}$, compared with 2×10^{-3} for V_2O_3—about the same ratio as the γ-values.† The enhancement, though not so large, should therefore be of Brinkman–Rice rather than Stoner type, as expected for a half-full e_g band.

Low-temperature NiS is sometimes called an "itinerant antiferromagnetic".

---

† Compare for instance metallic Pd, for which the degeneracy temperature deduced from C_v is 1750 K and that deduced from the Pauli susceptibility is 500 K.

7   Some metallic transitional-metal compounds with ferro- or antiferromagnetic order

7.1   Chromium dioxide (CrO_2)

This material has the rutile structure, and is metallic, the resistivity behaving like T^2 at low temperatures (Rodbell *et al.* 1966). It is ferromagnetic with a saturation moment of $2\mu_B$ per Cr atom, the maximum possible value for a $3d^2$ state with no orbital angular momentum.

The c/a ratio is greater than for VO_2, which implies that the π^* band (i.e. that with d-orbital lobes in the basal plane) is more occupied than in VO_2 (Goodenough 1971, p. 352). But we think that if it were not ferromagnetic, the π^* band, in contradistinction to VO_2, would be wholly above the Fermi energy. The Hubbard correlation term U, however, produces localized moments for the $3d^2$ states, as explained in Chapter 3, and these, if oriented ferromagnetically, would just fill the $t_\parallel$ band. The filled band (for spin-up electrons) will now overlap the π^* band, allowing ferromagnetic interaction of Zener or RKKY type between the d^2 moments, as described in Chapter 3. The T^2 term in the resistivity could be explained as in Chapter 2, Section 6.

Other ferromagnetic metallic oxides are CoS_2 and FeS (high-spin d^6). In these also the moment appears close to the value for the free ion (Wilson 1972).

7.2   Transitional-metal sulphides and selenides with pyrite structure of the form MS_2 and MSe_2; $Ni(Se_{1-x}S_x)_2$

In this structure the S_2 or Se_2 pair accepts two electrons, and the metallic ion is of the form M^{2+}. A wide variety of behaviour is observed. Thus FeS_2 is a semiconductor with a band gap equal to 0.9 eV; the Fe^{2+} $3d^6$ ion is in the low-spin state, the six electrons filling a t_{2g} band. The addition of CoS_2 provides donors, as few as 3% of Co ions producing a degenerate electron gas with conductivity about $250\,\Omega^{-1}\,cm^{-1}$ at low temperatures and *decreasing* as the temperature is raised (Jarrett *et al.* 1968). CoS_2, with *one* electron in the e_g band is metallic and ferromagnetic; Waki and Ogawa (1972) reported a moment of slightly less than μ_B, while Jarrett *et al.* showed that the moment increases uniformly from zero to about μ_B in the series $Fe_{1-x}Co_xS_2$. Waki and Ogawa found a large electronic linear specific heat ($\gamma = 9.5\,mJ\,K^{-2}$), which corresponds to a bandwidth of almost 0.5 eV. We have here a metallic correlated gas filling less than half of a d-band (e_g), and we have seen that this favours ferromagnetism. The Brinkman–Rice arguments for enhanced specific heat apply here too.

NiS_2, particularly its behaviour under pressure, has been discussed by Wilson and Pitt (1971) and Wilson (1985). This has two electrons *half* filling an e_g band, and should therefore behave like V_2O_3. It is an antiferromagnetic semiconductor, but shows a metal–insulator transition at a pressure of 46 kbar (Mori *et al.* 1973). Thus the transition occurs for a decrease in volume of about 0.4%, with no change

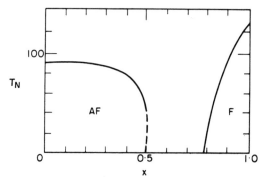

Fig. 6.18 Magnetic phase diagram of Co$(S_xSe_{1-x})_2$ (Adachi *et al.* 1970). See also Adachi *et al.* (1979).

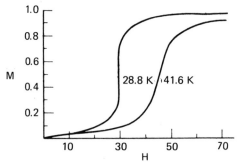

Fig. 6.19 The magnetic moment of Co$(S_xSe_{1-x})_2$ in the paramagnetic range as a function of field. Adapted from data given by Adachi *et al.* (1970, Figs. 5–7).

in crystal structure (Endo *et al.* 1973). It also becomes metallic on alloying with Ni$_2$Se$_2$, and the behaviour of this system is discussed later in this section.

Of particular interest is the series Co$(S_xSe_{1-x})_2$, the magnetic behaviour of which is shown in Fig. 6.18. The whole series is metallic. It should in our view be treated at the sulphur-rich end as a highly correlated gas with one electron per atom in an e$_g$ band. The condition for ferromagnetism is likely to be fulfilled for a degenerate band less than half full. The transition to paramagnetism is doubtless due to the broadening of the band due to admixture of 4p-orbitals from Se. The moment apparently changes discontinuously, not as $(x-x_0)^{1/2}$. This, we think, must be an example of the effect described in Chapter 3 (Fig. 3.10), the energy–moment curve having two minima, one at $M=0$ and one near $M=\mu_B$. This could occur because the admixture of 4p-orbitals required to minimize the energy will be small in the ferromagnetic state (one wants to minimize B), and large in the paramagnetic state (one now wants to minimize U). The argument is the same as for NiS in Section 6. The effect is shown well by the behaviour of M as a function of H found by Adachi *et al.* (1970) and shown in Fig. 6.19. The sudden increase of M with field must be due to an energy–field curve of the type shown in Fig. 3.10.

More surprising, perhaps, is the formation of an antiferromagnetic phase for Se-rich alloys, for which the bandwidth should be large and U, if anything,

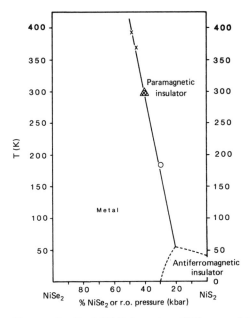

Fig. 6.20 Phase diagram for the Ni(S/Se)$_2$ system (Wilson and Pitt 1971). There is an approximate equivalence between Se doping and pressure: 1%Se $\equiv$ 1 kbar.

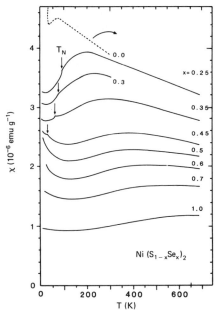

Fig. 6.21 Magnetic susceptibility for the system Ni(S/Se)$_2$. Note that the maximum in χ occurs well above the Néel point for the compositions $0.25 < x < 0.45$ (which are all metals at low temperature). From Ogawa (1979).

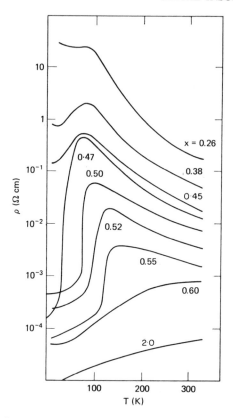

Fig. 6.22   Resistivity versus temperature for single crystals of $NiS_{2-x}Se_x$ (Bouchard *et al.* 1973).

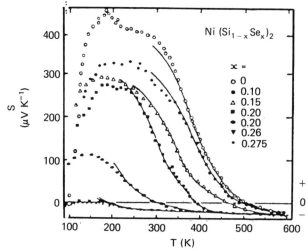

Fig. 6.23   Thermopower for $Ni(S/Se)_2$. From Kwizera *et al.* (1981).

diminished. We think that the crystal-field splitting is here becoming small compared with the bandwidth B, so that a gradual transition to the high-spin state occurs, which would increase the tendency to antiferromagnetism.

The series $Ni(S_{1-x}Se_x)_2$ has been investigated by Jarrett et al. (1973), Krill et al. (1976), Mabatah et al. (1980) and Kwizera et al. (1981), and a detailed discussion has been given by Wilson (1985). The series has the pyrite structure, which can be thought of as simple cubic (rock salt) like NiO with the oxygen anions replaced by the axial complex $(S—S)^{2-}$, $(S—Se)^{2-}$ or $(Se—Se)^{2-}$. Wilson describes this as the "ideal" metal–insulator transition, which occurs at low temperatures at $x = 0.3$ from the antiferromagnetic insulator (as in NiS_2) to an *antiferromagnetic* metal. We have seen (Chapter 4, Section 5) that transitions to an antiferromagnetic metal or to a highly correlated one are both possible. The Néel temperatures are of order 50 K and antiferromagnetism disappears at concentrations between $x = 0.45$ and $x = 0.5$. The properties are very sensitive to pressure; there is an approximate equivalence between Se concentrations and pressure, following $1\% \, Se \equiv 1 \, kbar$. A phase diagram due to Wilson is shown in Fig. 6.20 and the magnetic susceptibility in Fig. 6.21.

The resistivities of the mixed crystals are shown in Fig. 6.22. It can be seen that (as indeed follows from the phase diagram, Fig. 6.20) the transition to the *insulating* state occurs with larger x for increasing temperature, probably because of the high entropy of the AF insulator above its Néel temperature.

Clearly the metal–insulator transition is by no means so sharp as predicted for a first-order change of phase, but much disorder broadening is observed in the Raman spectra (Lemos et al. 1980).

Thermopower measurements due to Kwizera et al. (1981) are shown in Fig. 6.23. The equation for the thermopower in the metallic state, $S = (\frac{1}{3}\pi^2 k_B^2 T/e) \, d \ln \sigma/dE$, should not be valid above about 150 K; it can be seen that $d \ln \sigma/dE$ must be about 0.03 eV, indicating a very narrow band in the metallic state, possibly due to Brinkman–Rice enhancement or formation of spin polarons.

7

Some Metal–Insulator Transitions
in Various Materials

1 Introduction

We discuss here several materials in which a metal–insulator transition appears
to take place, probably of Anderson type, disorder playing a role. A feature of
these materials is that for many of them the preexponential factor seems to have
the value $\sigma_{\min}$ in the semiconducting range, and over a certain temperature range
in the metallic state the conductivity is also equal to $\sigma_{\min}$. We interpret this by
assuming that the interaction with phonons (or perhaps spin waves) is so strong
that in the Kawabata equation (Chapter 1, equation (52)) $L_i \sim a$, so that the
quantum interference effect is absent. We shall show in Chapter 10 that this is so
for liquids. Few of these materials have been investigated at very low
temperatures, however, and if they were then we should expect σ to tend to zero at
the transition, instead of showing an apparent minimum metallic conductivity.

In some of the metal–insulator transitions discussed here the use of classical
percolation theory has been used to describe the results. This will be valid if the
carrier cannot tunnel through the potential barriers produced by the random
internal field. This may be so for very heavy particles, such as dielectric or spin
polarons. A review of percolation theory is given by Kirkpatrick (1973). One
expects a conductivity behaving like

$$\sigma \sim \sigma_0 \left(\frac{x - x_0}{x_0} \right)^{\nu}, \quad \nu = 1.6,$$

where σ_0 in many materials is probably near σ_{IR} (Chapter 1, equation (38)).

2 Lanthanum–strontium vanadate ($La_{1-x}Sr_xVO_3$)

This material, which has the ideal cubic perovskite structure, has been extensively
investigated from both theoretical and experimental points of view (see Dougier

199

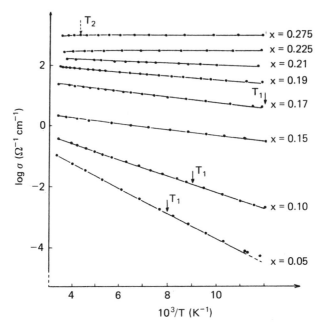

Fig. 7.1 Variation of electrical conductivity of $La_{1-x}Sr_xVO_3$ as a function of reciprocal temperature (Dougier and Hagenmuller 1975).

and Casalot 1970, Dougier and Hagenmuller 1985). It undergoes a transition from insulator to metal at about $x = 0.2$ (Fig. 7.1). For lower values of x the material is antiferromagnetic, the Néel temperature T_1 dropping with x from 150 K to zero at the transition. The transition has been judged to be of Anderson type, with a minimum metallic conductivity of about $200\,\Omega^{-1}\,cm^{-1}$ in both the preexponential factor and the metallic range down to a temperature of about 100 K. The disorder responsible is that due to the random positions of the Sr^{2+} ions. Mott and Davis (1979, p. 144) point out that for a normal 3d-band this random potential is not sufficient to produce Anderson localization well into the band, and some kind of mass enhancement must be assumed. This could arise from the presence of a degenerate gas of dielectric polarons. Another possible source of mass enhancement is the formation of spin polarons, which can occur because of the antiferromagnetic nature of the material. As we have seen in Chapter 2, either can lead to a mass enhancement.

Interaction either with phonons or spin waves might be responsible for the condition $L_i \sim a$ near the transition. At lower temperatures σ in the metallic range should drop continuously to zero at the transition; further observations are desirable to see if this is so.

Figure 7.2 shows the activation energy as a function of strontium concentration x, deduced from the results of Fig. 7.1. The linear behaviour followed by a tail is very similar to the plot of ε_2 in Fig. 5.2. The reason for the tail

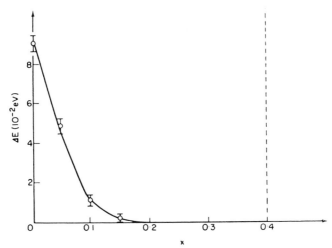

Fig. 7.2 Variation with x of the activation energy deduced from the results of Fig. 7.1.

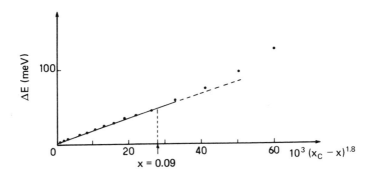

Fig. 7.3 Variation of the activation energy of the electrical conductivity of $La_{1-x}Sr_xVO_3$ as a function of $(x_c - x)^{1.8}$ (Dougier and Hagenmuller 1975).

is not understood. Figure 7.3 shows a variation in the tail as $(x - x_c)^v$, with $v = 1.8$. This again is not understood; we should expect $v = 1$. This value of the index suggests that classical percolation theory might be valid. This seems just possible if the particles are very heavy (for example as a consequence of the formation of spin polarons).

Figure 7.4 shows the thermopower S. For $x < 0.1$, S increases with decreasing temperature, suggesting excitation to a mobility edge; for $x = 0.15$ the reverse is the case, suggesting hopping. Variable-range hopping ($\log \rho \propto T^{-1/4}$) has been observed at low T by Sayer et al. (1975), who found, using the analysis of Chapter 1, Section 15, $\xi^{-1} \sim (n - n_c)^v$, with $v = 0.6$ instead of the value $v = 1$ expected, but the unknown effect of a Coulomb gap makes this result difficult to interpret.

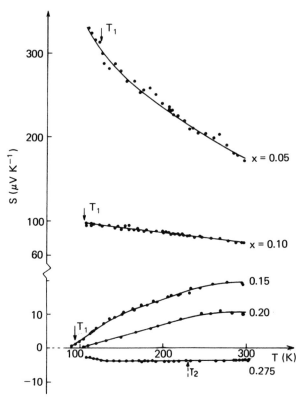

Fig. 7.4   Variation of the thermopower with temperature for $La_{1-x}Sr_xVO_3$ (Dougier and Hagenmuller 1975). T_1 is the Néel temperature.

3   Vanadium monoxide (VO_x)†

This material has the simple cubic structure. When $x = 1$ the three electrons in the V^{2+} ion will half fill the t_{2g} band. Even when stoichiometric ($x = 1$), it normally contains about 15% of vacancies of both signs. Figure 7.5 shows the thermopower S of VO_x at room temperature as a function of x, contrasted with that of TiO_x. We note that for both materials the low values of S indicate either metallic behaviour or hopping by electrons at the Fermi energy consequent on Anderson localization in the random field produced by the vacancies. The contrasting behaviour of the two materials shows, in our view, that the electrons in VO form a highly correlated gas of the kind postulated by Brinkman and Rice (1970 b), and described in Chapter 4, Section 6. There are thus two kinds of carrier in the metallic phase, essentially mobile V^{3+} and V^+ electronic configurations, which behave like electrons and holes. This possibility is mentioned in Chapter 4,

---

† The experimental data quoted in this section are from Banus and Reed (1970) and Goodenough (1971, 1972).

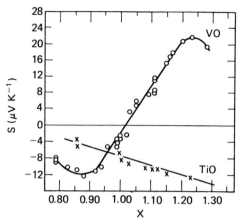

Fig. 7.5 Thermopower of VO_x and TiO_x at room temperature as a function of x (Banus and Reed 1970).

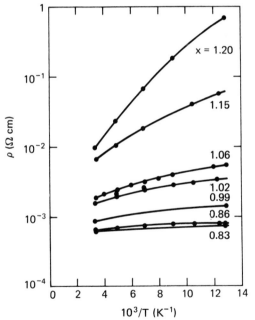

Fig. 7.6 Resistivity of VO_x as a function of reciprocal temperature for various values of x. From Banus and Reed (1970).

Section 6, but a mathematical treatment has not, as far as we know, been given. However the results shown in Fig. 7.5 make it very probable that it is so, if weak localization is present, though it appears unlikely without.

Figure 7.6 shows the resistivity plotted against $1/T$ for various values of x. A metal–insulator transition occurs with $x \approx 0.8$ and $\sigma_{min} = 2 \times 10^3 \, \Omega^{-1} \, cm^{-1}$, which is rather a large value—perhaps because the orbitals are d-like (see

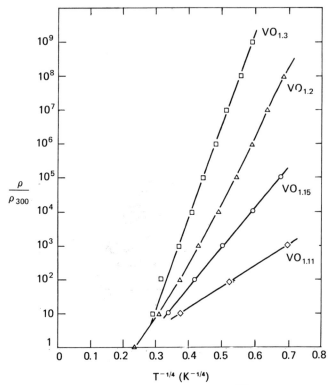

Fig. 7.7   Conductivity of VO_x as a function of $T^{-1/4}$ (Norwood and Fry 1970).

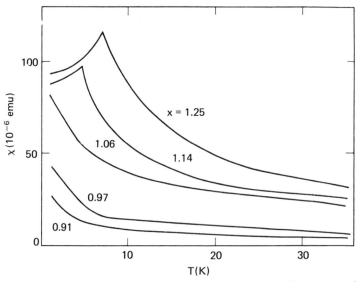

Fig. 7.8   Magnetic susceptibility of VO_x at low temperature (Kawano *et al.* 1966).

Chapter 1, Section 7). The fact that σ_{min} exists must mean that $L_i \sim a$, indicating very strong interactions with phonons or spins. It looks as if oxygen vacancies give weaker localization than those of vanadium; the reason why is not clear, but may be associated with the possibility that the latter disturb the d-band more.

Figure 7.7 shows that at low temperatures hopping conduction is observed, giving further evidence of weak localization.

Also of interest are the magnetic properties shown in Fig. 7.8. The high Pauli paramagnetism is to be expected, whether or not states are Anderson-localized, for a highly correlated gas. In addition, there appears to be a small number of free moments (7–15% of sites), which give rise to Curie paramagnetism with very low T_N (a few K). In our view, this must be a Kondo phenomenon, and we suggest that when two vanadium ions are not separated by an oxygen, or perhaps in some more complicated situations, a configuration is formed, with a moment, giving a high value of E_a (Chapter 3, Section 7) for the introduction of a conduction electron. E_a/Δ must be large; perhaps Δ for a highly correlated gas is small. According to Kawano et al. (1966), the susceptibility shows a Néel temperature. If this is correct, we can cite this as evidence that a random assembly of magnetic centres, with RKKY or other interaction between them, can show a sharp Néel temperature, rather than the properties of a spin glass.

4 Magnesium–bismuth amorphous films

Ferrier and Herrell (1969, 1970) and Sik and Ferrier (1974) measured the resistivity and thermopower of amorphous films of composition $Mg_{3-x}Bi_{2+x}$. The room-temperature conductivity as a function of bismuth concentration and the temperature coefficient of resistivity are shown in Figs. 7.9 and 7.10. The above authors and Mott and Davis (1979, p. 130) have interpreted these results in terms of the pseudogap model of Chapter 1, Section 16. This was criticised by Sutton (1975) on the basis of optical measurements; he found that Mg_3Bi_2 is a semiconductor with a gap of 0.15 eV. We therefore draw the density of states as in Fig. 7.11. The band formed by Bi^{3+} (which is full) and that by Mg^{2+} (which is empty) do not overlap. Addition of small amounts of Bi or Mg, however, is unlikely to change the density of states in any major way, or produce donors, acceptors or impurity bands; they will simply introduce carriers into one or other band tail, producing there a degenerate electron gas of electrons or holes, a metal–insulator transition occurring when $E_F = E_c$. The thermopower, shown in Fig. 7.12, increases with temperature, suggesting that conduction in this regime is by hopping rather than by excitation to a mobility edge.

However, the gap may in fact be greater than suggested by Sutton (Andrew Long, 1988 private communication, based on discussions in 1977 with M. Sik). Sik's films were much thicker than those of Sutton, and he estimates the true gap to be about 0.7 eV. He considers that in Sutton's work fluctuations in composition may fill in the gap. More recent unpublished work (Odeh 1978) optically measured an energy gap of 0.264 ± 0.006 eV in a film of $60:40$

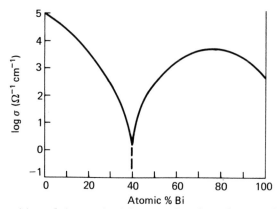

Fig. 7.9 Logarithm of the conductivity as a function of composition of amorphous evaporated films of Mg–Bi. The deposition was at 80 K (Herrell 1970) and the measurements were made at this temperature. From Ferrier and Herrell (1969).

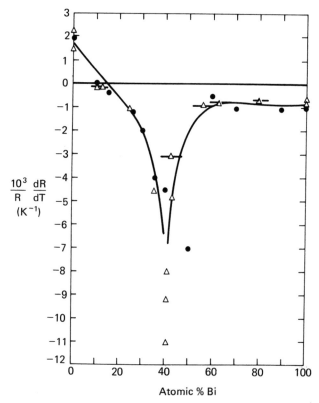

Fig. 7.10 Temperature coefficient of the resistivity of amorphous Mg–Bi. From Ferrier and Herrell (1969).

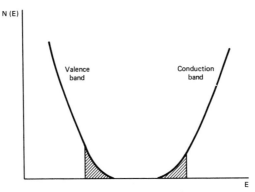

Fig. 7.11 Density of states for amorphous Mg–Bi, rigid-band model. The gap may be 0.4–0.6 eV after annealing.

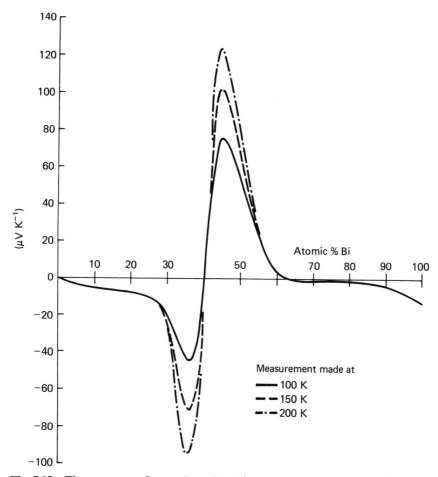

Fig. 7.12 Thermopower of amorphous Mg–Bi. From Ferrier and Herrell (1969).

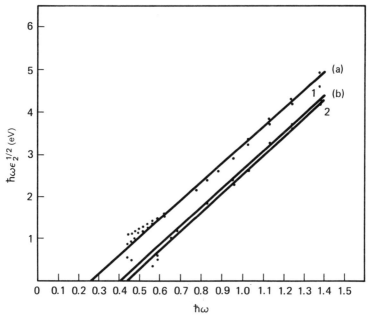

Fig. 7.13 $\hbar\omega\varepsilon_2^{1/2}$ as a function of photon energy $\hbar\omega$ for a film of amorphous Mg–Bi with 40.4 ± 1.6 at.% Bi: (a) as deposited at 85 K; (b) annealed at 200 K. 1 and 2 refer to two positions of the optical sample. From Odeh (1978).

composition as deposited (temperature 85 K when measured), which increased to 0.42 ± 0.02 eV on annealing. It was concluded that there are some microscopic defects or inhomogeneities, which can be annealed away close to the crystallization temperature. These results are reproduced in Fig. 7.13.

5 Tungsten bronzes

The bronzes with composition M_xWO_3, where M is an alkali metal, show metallic behaviour, with conductivities of about $500\,\Omega^{-1}\,cm^{-1}$ near the transition (Lightsey 1973), rising to 10^4–$10^5\,\Omega^{-1}\,cm^{-1}$ for higher values of x. The metal–insulator transition is associated with one or more changes of phase to different structures with insulating properties; for details see Hagenmuller (1971) and Doumerc *et al.* (1985). This can, however, be avoided by the addition of fluorine or tantalum (see below).

   In the metallic regime the electronic specific heat (Vest *et al.* 1958, Zumsteg 1976) and the magnetic susceptibility (Zumsteg 1976, Graener *et al.* 1969) provide evidence that the effective mass should be given by $m_{\text{eff}}/m \approx 1.6$, but the measured quantities vary as x rather than $x^{1/3}$, as would be the case for a parabolic band form. Tunstall (1975) suggested that the band has a tail resulting from disorder; if $N(E) = C e^{-E/E_0}$ then

$$n = \int N(E) \, dE = C E_0 e^{-E/E_0},$$

so that $N(E)$ is proportional to n, and thus to x.

Mackintosh (1963) showed that the above effective mass, and an appropriate dielectric constant, are comparable with a Mott transition according to the equation $n^{1/3} a_H = 0.25$ at $x = 0.1$, but this equation has no theoretical basis in an exponential band tail.

Various authors (Fuchs 1965, Lightsey 1973, Webman et al. 1976) have discussed the conductivity in terms of classical percolation theory, but we think this is unlikely to be correct for particles for which there is no evidence for mass enhancement.

In $Na_x WO_{3-y} F_y$, Doumerc (1978) observed a transition that has all the characteristics of an Anderson transition; similar phenomena are observed in $Na_x Ta_y W_{3-y} O_3$. The results are shown in Fig. 7.14. It is unlikely that this transition is generated by the overlap of two Hubbard bands with tails (Chapter 1, Section 4); this could only occur if it took place in an uncompensated alkali-metal impurity band, which seems inconsistent with the comparatively small electron mass. We think rather that in the tungsten (or tungsten–tantalum) 5d-band an Anderson transition caused by the random positions of Na (and F or Ta) atoms occurs. The apparent occurrence of σ_{min} must, as explained elsewhere, indicate that $L_i \sim a$ at the temperature of the experiments. Work below 100 K, to look for quantum interference effects, does not seem to have been carried out.

Hollinger et al. (1985) have studied bronzes $Na_x WO_3$ and $Na_2 Ta_y W_{1-y} O_3$ near the metal–insulator transition using photoelectron spectroscopy with synchrotron radiation. The results show that the transition is due to localization in an impurity band in a pseudogap.

6 Metal–rare-gas systems

Metal–insulator transitions in frozen mixtures of metals and rare gases have been extensively investigated over a number of years, partly in the hope of finding a discontinuous Mott transition if the metal has an odd number of outer electrons. Most observations, however, are now interpreted on the assumption that the mixtures have a granular structure and that the observed metal–insulator transition can be described by classical percolation theory. A summary is given by Micklitz (1985). The conductivity near the transition behaves like $(x - x_c)^\nu$, with ν in the range 1.5–2.0 expected from percolation theory; the predicted value is 1.6 (Kirkpatrick 1973). An example is Sn–Ar (Ludwig et al. 1981), which becomes insulating at $x = 0.32$ and for which $\nu = 1.6 \pm 0.1$. At present the only system believed to show a random mixture on an atomic scale of metal and gas atoms is Bi–Kr (Ludwig and Micklitz 1984). The firmest evidence is that the superconducting transition temperature T_c shows a strong decrease with decreasing concentration of bismuth, which is not the case for Sn–Ar, which shows that in the latter material percolation paths with constant T_c exist in the

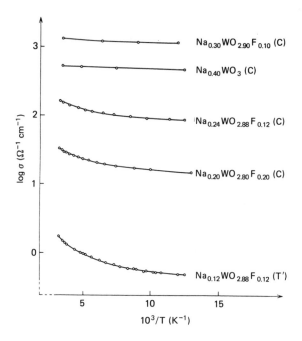

Fig. 7.14 Logarithm of conductivity as a function of reciprocal temperature for $Na_xWO_{3-y}F_y$.

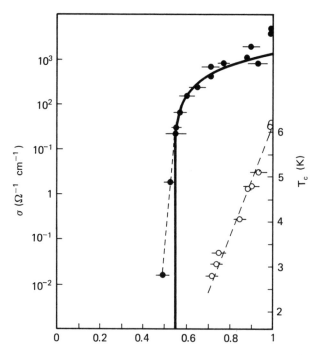

Fig. 7.15 Conductivity σ (solid symbols) and superconductivity transition temperature T_c (open symbols) in Bi–Kr as functions of atomic bismuth concentration x. The solid line is a fit to the equation $\sigma = const \times (x - x_c)^\nu$ with $x_c = 0.55$ and $\nu = 1.07$. The dashed lines are simply guides to the eye.

whole "metallic" region. For Bi–Kr the observed value of v is 1.07 ± 0.1, compatible with the theoretical value $v = 1$. Ludwig *et al.* point out that deposited bismuth films are known to be amorphous, while those of tin are not, which doubtless accounts for the different behaviour.

Figure 7.15 shows the conductivity in the limit of low T for Bi_xKr_{1-x}, together with the superconducting transition temperature T_c. The rapid drop when σ falls below σ_{min} ($\sim 300 \, \Omega^{-1} \, cm^{-1}$) is striking, the linear drop taking place in $\delta x/x_c \approx 0.2$, in contrast with $\delta x/x_c \approx 1$ in a-NbSi. Since the number of outer electrons in Bi is odd and there can be no compensation, the question arises as to why a discontinuous Mott transition is not observed. We think that this is in principle a Mott transition, like that of Na_xNH_3, and that the region near the transition is unstable against separation into two phases; this may be why it apparently occurs in most metal–rare-gas films. We have here, then, a transition in which two Hubbard bands, with localized tails due to disorder, broaden until $E_F = E_c$. If this is correct then some way from the transition we may expect the system to be a spin glass, with zero value of $N(E_F)$, and no variable-range hopping. Further observations would be of interest.

8

Wigner and Verwey Transitions

1 Wigner crystallization

The metal–insulator transitions described in the preceding chapters occur for electrons in a narrow band, for which the orbitals can be described by the tight-binding method as made up of atomic wave functions. More than a decade before this kind of transition was considered, Wigner (1938) proposed that for low densities a gas of free electrons at zero temperature would crystallize into a non-conducting state, even in the absence of any lattice field. The argument is as follows. We suppose that there are n electrons per unit volume neutralized by a background of uniform positive charge ("jellium"). Then, if we localize each of them in a volume n^{-1}, this costs an amount of energy, per electron,

$$\sim \hbar^2 n^{2/3}/m. \tag{1}$$

On the other hand, there is a gain of interaction energy between electrons

$$\sim e^2 n^{1/3}. \tag{2}$$

At low densities, then, energy is always to be gained by such localization, which should take place when

$$n^{1/3} a_H = \text{constant} \qquad (a_H = \hbar^2/me^2)$$

The evaluation of the constant, since it must compare the energies of a low-density electron gas with the crystallized state, is difficult. An estimate (Mott 1961, Kugler 1969) is 0.05.

The concept of a degenerate electron gas superimposed on a uniform background of positive charge is of course difficult to realize. The present author (Mott 1967) suggested that a good approximation to this situation would be a highly doped and very strongly compensated semiconductor, for which the Bohr radius $\kappa \hbar^2/me^2$ was large compared with the distance between centres. This idea was developed by Durkan et al. (1968); and Somerford (1971) used strong magnetic fields to induce such a transition in n-type InSb; Care and March (1971) and Rousseau et al. (1970) gave evidence to show that conduction in such a lattice is due to thermally generated vacancies.

Our conjecture as to the properties of a Wigner crystal are as follows.

(i) It will be non-metallic and antiferromagnetic (or perhaps ferromagnetic, cf. Edwards and Hillel 1968).

(ii) There will be two conduction mechanisms:

   (a) excitation of electrons to the "conduction band"—a process similar to that described earlier in this chapter for "Mott–Hubbard" insulators; the material should be an intrinsic semiconductor and both electrons and holes can carry current;

   (b) formation of vacancies, which are mobile, as described by Care and March (1971); the vacancies are identical with the holes mentioned in (a), but, if they are formed, the gas (if held at constant volume) undergoes compression.

(iii) As the temperature is raised, the lattice can "melt"–by which we mean that long-range order will disappear.

Wigner crystallization is a subject on which there is little theoretical work, and it is not certain whether the phenomenon has been observed.

Experimental data on $Hg_{1-x}Cd_xTe$, a material in which for appropriate values of x the gap disappears, have been claimed as showing evidence for such a phenomenon (Nimtz et al. 1979, Rosenbaum et al. 1985).

Whether or not the available sites for electrons lie on the crystal lattice, the electrons may crystallize under the influence of their mutual repulsion. At low temperatures the system will then be non-conducting. There will be an energy of order $e^2/\kappa a$ needed to transfer an electron to a "wrong" site, so the material should behave like an intrinsic semiconductor. However, this energy should decrease as more and more electrons are transferred, so that as the temperature is raised we expect a first- or second-order transition to a state in which the sites are occupied in a random way. The analysis of such a transition could be analogous to that given originally by Bragg and Williams (1934) (for references to early work see Mott and Jones 1936, p. 29) for the order–disorder transition in alloys of the types AB, AB_3, etc. We call such transitions "Verwey transitions", since they were first proposed by Verwey (1935) for Fe_3O_4.

Such transitions are expected particularly for transitional-metal oxides of mixed valence, where polaron formation increases the effective mass of the carriers, so that the kinetic energy is small compared with the potential energy. The Hubbard intra-atomic energy U (Chapter 2, Section 9), of order $\langle e^2/r_{12} \rangle$ and equal to several electron volts, must be distinguished from the much smaller energy $\sim e^2/\kappa a$ needed to move an electron (polaron) onto a "wrong" site. This chapter is devoted to the discussion of materials in which the number of sites, N, is greater than n and in which, because of polaron formation or for other reasons, $\hbar^2/m_p a^2$ is much smaller than U, so that an essentially localized model is a useful first approximation, though $\hbar^2/m_p a^2$ and $e^2/\kappa a$ may be comparable (here m_p is the polaron effective mass).

We suppose that we have a gas of n electrons moving between N equivalent sites, and $N > n$. Then we have to ask whether or not such a material will conduct at low temperatures. There are several possibilities.

(i) It will form a degenerate electron gas and conduction will be of the kind usual in metals.

(ii) If the N sites are distributed at random or if there is some other kind of disorder then Anderson localization may occur at the Fermi energy, so that conduction at low temperatures will be by hopping (Chapter 1, Section 15). This can be the case under certain circumstances for doped and compensated semiconductors for low concentrations and for $La_{1-x}Sr_xVO_3$ (Chapter 7) and the oxide superconductors (Chapter 9).

The materials to be discussed in this chapter are Fe_3O_4 (magnetite) and Ti_4O_7. In both, the number of carriers is half the number of available sites. At low temperatures they take up ordered positions, under the influence of their mutual repulsion; in both, there is a first-order transition as the temperature is raised to a state in which the carriers are at random sites. In Ti_4O_7 the carriers are bipolarons, as has already been pointed out in Chapter 2.

2 Fe_3O_4 (magnetite)

A recent review of the experimental situation has been given by Honig (1985). It is pointed out that the electrical properties, particularly near to the transition, are very sensitive to purity and specimen preparation, and that much of the extensive experimental work is therefore open to doubt. None the less, the broad features of the behaviour of this material are clear. The history of the so-called Verwey transition in this material goes back to 1926, when Parks and Kelly (1926) detected an anomalous peak near 120 K in the heat capacity of a natural crystal of magnetite. The first detailed investigations were those of Verwey and co-workers (Verwey 1939, Verwey and Haayman 1941, Verwey et al. 1947), who showed that there was a near discontinuity in the conductivity at about 160 K. The conductivity as measured by Miles et al. (1957) is shown in Fig. 8.1.

In this spinel, to which we may give the formula $Fe^{3+}(Fe^{2+} + Fe^{3+})O_4$, below 119 K the Fe^{2+} and Fe^{3+} ions are ordered onto alternate (011) B-site layers, distorting the crystal into orthorhombic symmetry (Goodenough 1963, p. 185). It will be seen from Fig. 8.1 that there is a first-order transition at 119 K (the Verwey temperature T_V) and that above this the conductivity increases by a further factor ~ 10 and eventually decreases slightly. Its largest value ($\sim 10^2\,\Omega^{-1}\,cm^{-1}$) is rather low for a normal metal, however short the mean free path. In the ordered phase at the lowest temperatures the observed activation energies are small ($\sim 0.03\,eV$), and conduction is probably by a hopping mechanism, allowed on account of vacancies or other defects. At temperatures below 850 K the ions on A- and B-sites are ordered antiferromagnetically, so the current carriers (on the B-sites) have moments ordered ferromagnetically above and below T_V.

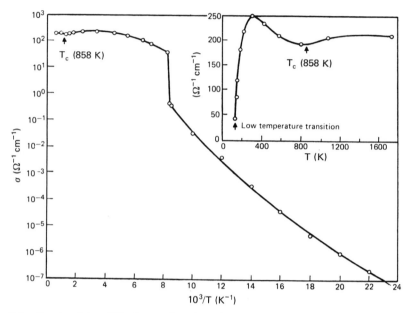

Fig. 8.1   Conductivity of Fe_3O_4 as a function of reciprocal temperature. After Miles *et al.*
(1957).

As explained in Section 1, we believe in common with most other
investigators that the Hubbard U can be considered large compared with the
bandwidth,† so that the conductivity is due to the fact that the number of carriers
(Fe^{2+}) is smaller (by a factor 2) than the number of the B-sites, on which there is a
mixture of Fe^{2+} and Fe^{3+}. Thus the quantity that determines the electrical
properties must be the ratio between the interaction between electrons (Fe^{2+} and
Fe^{3+}), which we write as $V = e^2/\kappa R$, and the bandwidth B. We write $V_0 = e^2/\kappa a$ for
the interactions between carriers on nearest-neighbour sites. On general grounds
we should guess that if V_0/B is small then the material will be a metal, and if it is
large then a Wigner crystallization will occur. An analysis of the condition for
crystallization was given by Lorenz and Ihle (1972) and Cullen and Callen (1971,
1973). The latter authors found that an ordered insulator will occur when
$V_0/B \gtrsim 3$. They also predicted multiple ordering for somewhat lower values of
V_0/B; such ordering is in fact observed in the insulating phase, there being much
evidence from Mössbauer and other measurements discussed by these authors. It
therefore seems right to conclude that V_0/B does not greatly exceed the critical
value.

---

† It is likely that the band is formed from non-degenerate orbitals, in which case, since all spins are
parallel, two electrons could not be on the same Fe atom even if U were not large.

We now ask what values of V_0 and B are likely. Bandwidths in transitional-metal oxides are 1–2 eV; the high magnetic-ordering temperature would not lead us to expect a small value of B, if it is given by a term of the type B^2/U (Chapter 3, equation (7)). We think that a value of V_0 of order 5 eV is very unlikely, in view of the small activation energies involved. It seems much more plausible that the carriers are small (or intermediate) polarons, so that

$$V_0 = \frac{e^2}{\kappa a},$$

with κ the background *static* dielectric constant, and B the polaron bandwidth, perhaps ~ 0.1 eV.

It is clear, however, that the activation energy for conduction at low temperatures, and the ordering energy, cannot be wholly electrostatic, since V_0/B lies not too far from the critical value, and the kinetic energy B must be taken into account. This follows particularly from the considerations of Anderson (1956), who pointed out that, if nearest-neighbour order is maintained, the electrostatic energy needed to produce long-range disorder is less than 10% of that to produce complete disorder; he also showed that half the entropy ($nk_B \ln 2$) for complete disorder is produced by disorder in which short-range order is maintained. The σ–$1/T$ curve shown in Fig. 8.1 suggests that it is necessary only to double the temperature after the transition to obtain a maximum value of the conductivity, and thus presumably complete disorder. Therefore no simple electrostatic model will be sufficient (cf. Fazekas 1972, Sokoloff 1972).

The assumption that the carriers are small or intermediate polarons in no way militates against discussions of the band structure of the ground state (see e.g. Camphausen *et al.* 1972, Cullen and Callen 1971, 1973). The absence of Jahn–Teller distortion (Goodenough 1971) also, in our view, indicates *not* the absence of a polaron mass-enhancement but rather a value of V_0/B not too far from the critical value. These conclusions seem to be in agreement with the considerations of Sokoloff (1972), who used a description in terms of a degenerate band of small polarons. Samara (1968) showed that pressure lowers the temperature of the Verwey transition. If this depended only on $e^2/\kappa a$ then the opposite should be the case. But pressure will increase B, and push the substance nearer to the critical value for the metal–insulator transition.

Miyahara (1972) showed that most impurities, including a deficit of metal, lower T for the metal–insulator transition. More recently, Whall and co-workers (see Graener *et al.* 1979) showed that 5% of fluorine in $Fe_3O_{4-x}F_x$ destroys the sharpness of the transition;[†] their results are shown in Fig. 8.2, where $\log \sigma$ is plotted versus $1/T$. The straight line in the insulating range does not necessarily mean that conduction is by hopping; it is suggested that, near the transition, conduction is by excitation of electrons from the Fe^{2+} sites into the Fe^{3+} sites, from which they can move by band motion, whether or not they form polarons.

[†] Later work by Barlow *et al.* (1987) shows that traces of the sharp transitions, on doping with F or Ge, remain at higher concentrations.

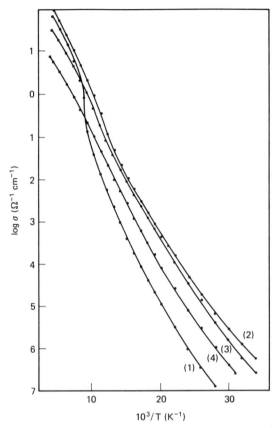

Fig. 8.2 Conductivity of $Fe_3O_{4-x}F_x$ for various values of x as a function of reciprocal temperature (Whall *et al.* 1978): (1) $x=0.025$; (2) 0.05; (3) 0.1; (4) 0.15. It will be seen that the Verwey transition disappears somewhere between $x=0.025$ and 0.05.

The disappearance of the sharp Verwey transition was discussed by Mott (1979), who suggested that at low temperatures the material is a "Wigner glass", the electrons (Fe^{2+} ions) being frozen into random sites and the whole system stabilized by the fluorine. Discussion of the thermopower measurements show, according to Mott (1979), that a hopping mechanism is operative at low T. Ihle and Lorenz (1985), however, consider that the electrons in the "wrong" sites move by a small polaron band mechanism.

The mechanism for conduction at temperatures above the transition has long been a matter of controversy. It is very low for a metal, less than $300\,\Omega^{-1}\,cm^{-1}$, of the order of σ_{min}. The temperature dependence is small (see Fig. 8.1). Ihle and Lorenz (1980, 1985) pointed out that traditional one-particle concepts break down because there is strong short-range order due to Coulomb repulsion between heavy charge carriers (perhaps intermediate polarons, see Chapter 2, Section 2). Mott (1979) proposed that the high-temperature phase in the pure material can be considered as a Wigner glass (like the impure material below T_c), stabilized by the entropy of disorder. The conductivity would be expected to be low, but to calculate it is difficult. Ihle and Lorenz (1985 and earlier papers) attempt this by combining hopping and band conduction.

3 Ti$_4$O$_7$

In this material also the sharp Verwey transition disappears with traces of impurity—in $(Ti_{1-x}V_x)_4O_4$ when $x > 0.01$ (Schlenker *et al.* 1976, Gourmala *et al.* 1978, Ahmed *et al.* 1978). Below 90 K the conductivity then behaves like $\exp[-(T_0/T)^{1/4}]$. The above authors believe that the conduction process giving these low activation energies is the hopping of an electron from one V–Ti^{3+} pair to a V–Ti^{4+} pair, there being some compensation to make this possible.

Mott (1979, p. 357) discusses the amount of impurity needed to make the sharp Verwey transition disappear. The small value, $x \sim 0.01$, for Ti$_4$O$_7$ is difficult to explain.

9

High-Temperature Superconductors and the Metal–Insulator Transition

1 Introduction

In this chapter we do not attempt to present a full theory of high-temperature superconductors: at the time of writing both experimental and theoretical papers fill the journals, and any such attempt would be out of date before this book could appear in print. Our objective is more modest. The original material investigated by Bednorz and Müller (1986), La_2CuO_4, has turned out to be an antiferromagnetic insulator, with a quasi-two-dimensional structure, the copper ions being in the state $3d^9$ (Cu^{2+}) and thus with a single positive hole and magnetic moment $\frac{1}{2}\hbar$. On doping with Sr or Ba, a further hole is introduced; this is now thought to be in the oxygen s–p band (see e.g. Emery 1987) (doubtless hybridized with Cu 3d-functions), so that no Cu^{3+} ions are formed. These holes will initially be bound to the *negative* charge resulting from the substitution of Sr^{2+} for a La^{3+}. A small level of doping results in the disappearance of antiferromagnetic order (but not, in our view, of the moments), the material showing spin-glass behaviour and remaining non-conducting. With further doping, a "metal–insulator" transition occurs, the metal being superconducting. Our aim in this chapter is to examine the nature of this transition and to propose, following other authors (Cyrot 1987, Kamimura *et al.* 1988, Su and Chen 1988), the formation of spin polarons and the possibility that they combine to form bipolarons, which are bosons and could replace the Cooper pairs of the normal metallic superconductors. We also discuss the effect of a two-dimensional structure.

In other high-temperature superconductors (e.g. $YBa_2Cu_3O_{7-x}$) the yttrium, barium and copper may be thought of as introducing $3+4+6=13$ electrons, and with $x=\frac{1}{2}$ the oxygen can receive this number. With $x>\frac{1}{2}$ there are holes in the oxygen band. So here again a metal–insulator transition of the same kind can take place.

In Chapter 7 we discussed the system $La_{1-x}Sr_xVO_3$, which for $x=0$ is an antiferromagnetic insulator and with increasing x undergoes a transition of Anderson type to metallic behaviour, with a decrease and ultimate disappearance of the Néel temperature. As far as is known, these materials are not superconductors. They do not have a two-dimensional structure.

Whatever the mechanism binding the two fermions together to form a boson, the question arises whether at zero temperature all the current carriers form bosons or, as in the BCS theory, only a small proportion of them in states near the Fermi energy. Only in the latter case will an equation for T_c of the type derived from BCS theory, $k_B T_c \sim \hbar \omega \, e^{-2/N(E_F)V}$, be valid. The observed small correlation length has led several investigators to assume that *all* the carriers form bosons (e.g. Sawatzky (1989) in his summary of a conference on the subject). In this case the transition temperature T_c is that at which the boson gas becomes completely non-degenerate; according to Prelovsek *et al.* (1987) this is given by

$$k_B T_c = 3.3 \hbar^2 N^{2/3}/m_{eff},$$

where N is the number of carriers per unit volume and

$$m_{eff} = (m_1 m_2 m_3)^{1/3},$$

m_1 etc. being the effective masses in the three directions. From the observed transition temperature Prelovsek *et al.* estimate $m_{eff} \sim 30 m_e$ for the bipolaron. At the time of writing we do not have a reliable estimate of the mass of the single polaron, which should be smaller; measurements for x outside the superconducting region are needed.

The proposal has been made that, if this model is correct, the carriers *above* T_c form a non-degenerate gas of bosons. Mott (1990) has pointed out that $d\rho/dT$ for ranges of x for which these materials are superconducting is greater than for larger values of x, and that very high values are reached for $T \gtrsim 300$ K. He also suggests that a spin bipolaron in a non-degenerate gas is heavier than in the partially degenerate superconducting state; measurements by Reagor *et al.* (1989) of the mass of the free carrier in undoped La_2CuO_4, presumably a bipolaron, is $m_{eff} \sim 1000 m_e$. The present author (Mott 1990) uses the model of spin polarons to show that above T_c the non-degenerate gas of bosons has a diffusion coefficient D independent of T, and thus by Einstein's equation $\sigma = ne^2 D/k_B T$, the resistivity must be linear in T.

He also uses Heikes formula for the thermopower (Heikes and Ure 1961),

$$S = \frac{k_B}{2e} \ln \frac{1-z}{z},$$

where z is the ratio of bosons to sites, and argues that for bosons z is near $\frac{1}{2}$, giving a small and temperature-independent value of the thermopower, observed in several experiments.

2   The metal–insulator transition in a doped antiferromagnetic insulator

We do not think that the transitions in $La_{2-x}Sr_xCuO_4$ or $La_{1-x}Sr_xVO_3$ can be "Mott transitions" as described in Chapters 4 and 6. Such transitions can take place on doping, as in the case of V_2O_3 doped with Ti_2O_3, but, as Wilson (1985) has emphasized, that can only occur when the free energies of the metallic and the antiferromagnetic (or RVB) states are very close. The transition should be first-order. The intermediate spin-glass region is not expected for this kind of transition. We think, then, that the transition is of Anderson type, similar in some respects to that which occurs in silicon doped with boron.

In the latter the insulating state near the transition is not antiferromagnetic; it has the nature of a spin glass, because of the high degree of disorder resulting from the random positions of the acceptors or donors. We have, however, argued that the carriers, whether in n- or p-type material, are in some sense spin polarons, this having the effect of *reducing* the effective mass (Chapter 5, Section 2), instead of increasing it as occurs in $Gd_{1-x}v_xS$ (Chapter 3, Section 5), the material that gives the strongest evidence for the existence of spin polarons. We think it likely that, when an atom of Sr replaces La in $LaVO_3$ or La_2CuO_4, a spin polaron is formed, which in this case enhances the effective mass. In the latter case, the spin polaron consists of a hole in the oxygen 2p-band polarizing some surrounding spins on Cu^{2+} ions parallel or more probably antiparallel to itself.[†] The polaron, being charged, is trapped by the negative charge on the dopant (Sr or Ba) acting as an acceptor. Clearly, this process must weaken the overall antiferromagnetic coupling; the antiferromagnetic order is destroyed, the material forming a spin glass, *before* the concentration reaches that for a metal–insulator transition. (Nothing similar occurs in Si:B, the non-metallic material already being a spin glass.)

We now consider the nature of the transition. In compensated Si:P the transition takes place in an impurity band; for high concentrations of dopant this merges with the conduction band. In uncompensated Si:P the many-valley structure of the conduction band leads to a kind of self-compensation so that $N(E_F)$ is already finite at the transition (Chapter 5), and the transition is of Anderson type. Whether this is so for p-type material or for single-valley materials is not known. If not then the transition must be of Mott type (Chapter 4), occurring when $B \sim U$.

In the case discussed here a Mott transition is unlikely; the Hubbard U deduced from the Néel temperature is not relevant if the carriers are in the s–p oxygen band, but if the carriers have their mass enhanced by spin-polaron formation then the condition $B \sim U$ for a Mott transition seems improbable. In those materials no compensation is expected. We suppose, then, that the metallic behaviour does not occur until the impurity band has merged with the valence band. The transition will then be of Anderson type, occurring when the random potential resulting from the dopants is no longer sufficient to produce localization at the Fermi energy.

---

† For a discussion of the conditions under which this is so see Eskes and Sawatzky (1989).

We have no direct evidence for the formation of spin polarons in any conductor, apart from gadolinium sulphide. The best evidence would be a decrease in the effective mass and increase in the conductivity of non-superconductors with magnetic field.

3 Magnetic bipolarons

As stated in Section 1, several authors have suggested that in high-temperature superconductors the following processes occur:

  (i) magnetic polarons form;

 (ii) they combine together to form bipolarons;

(iii) these are bosons, which give rise to superconductivity, replacing the phonon-generated Cooper pairs of the BCS theory, with a very small correlation length, estimated to be about 20Å for conduction in the planes.

The main evidence for bipolarons of the normal non-magnetic type comes from the properties of Ti_4O_7, where in a certain range of temperature carriers have no magnetic moment (Chapter 2, Section 3). For superconductivity one has to postulate a *degenerate* gas of bipolarons; (in Chapter 2 we introduced a degenerate gas of polarons). Such a model is of course an approximation, and we do not wish to assert that it is necessarily superior to other ways of describing interaction involving spins, of which there are many in the literature.

At the time of writing one of the most detailed descriptions of spin bipolarons is that of Su and Chen (1988). They discuss the formation of polarons in which the carriers have parallel spins, with the bipolaron having any moment from zero upwards, and alternatively in which they have antiparallel spins, with the bipolaron then necessarily having zero spin itself. In the former case the binding energy is of the same type as in dielectric bipolarons, varying as the square of the polarization—so when it is doubled through the overlap of two clouds, the energy per unit volume is multiplied by four, while the volume is perhaps halved. However, there are difficulties in this concept; bigger clusters might be formed, and the moment might depend on magnetic fields. Su and Chen, through numerical work, show that bonding is also possible when the carriers have antiparallel spins, when U/t is large, t being the overlap integral in the copper band. Their results are plotted in Fig. 9.1. They conclude that attraction can occur between two antiparallel-spin polarons if t is less than $\frac{1}{4}U$, apart from a repulsive tail. They also remark that the polaron will only be stable if this attractive force is greater than the Coulomb repulsion, which in its turn depends on the background dielectric constant.

The form of the spin polaron has been calculated, following earlier work for small U, by Schrieffer *et al.* (1988). It is described as having a cigar shape, lying in the CuO_2 planes, as shown in Fig. 9.2. Schrieffer *et al.* remark that any attractive

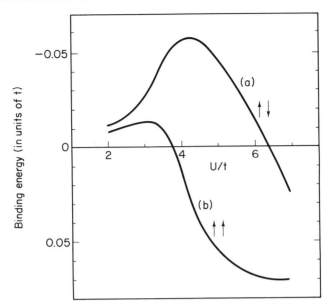

Fig. 9.1 Binding energy of a pair of spin polarons with (a) opposite and (b) parallel spins. From Su and Chen (1988).

force in two dimensions will produce a bound state, which is not the case in three dimensions, so the cigar form is very favourable for producing a bound pair. This, they suggest, is why two-dimensional structures are favoured. In any case, two dimensions favour localization.

In the superconductor Schrieffer *et al.* take the transition temperature to be

$$k_B T_c \sim E_F e^{-1/\lambda},$$

where λ is a coupling constant, given by

$$\lambda = b/\varepsilon_F,$$

with b the binding energy of a polaron.

If the concept of spin polarons as described here is correct, the most important theoretical problem at the present time is to describe the way in which they are bound together to form bosons, of dimensions about 20 Å and with binding energy of the order of the Néel temperature. Any binding mechanism will certainly distort the surrounding lattice, so the cohesion may well involve dielectric or dynamic Jahn–Teller terms, subject of course to the presence of only a very small (or zero) isotope effect.

We should also add that as long ago as 1985 Alexandrov and co-workers (see Alexandrov *et al.* 1986) were considering the production of superconductivity through the formation of dielectric bipolarons, but did not obtain high values of T_c, which are, however, predicted by Emin and Hillery (1989) in their work on "large" (dielectric) bipolarons.

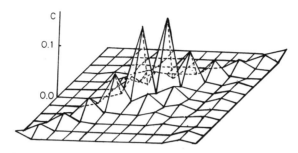

Fig. 9.2   Profile of charge deposited in a cigar-shaped polaron, according to Su and Chen
(1988).

4   Two-dimensional systems

In this section we discuss the electrical properties of materials that can be
envisaged as sheets of material with metallic conduction, separated by some
distance from each other. We suppose first that there is no disorder.

We denote by $\tau_\parallel$ the time of relaxation for conduction in the planes and by I
the transfer integral from one plane to another (this is often denoted by t). Then,
following Friend and Yoffe (1987), we ask whether $\tau_\parallel$ is greater or less than $\hbar/I$.

If $\tau_\parallel I/\hbar > 1$ then the system should form an anisotropic Fermi surface, with a
large effective mass perpendicular to the planes. In both directions we should
expect the same τ (namely $\tau_\parallel$), and the conductivities in the two directions should
be $ne^2\tau/m_1$ and $ne^2\tau/m_2$, where m_1 and m_2 are the effective electron masses.

If, on the other hand, $\tau_\parallel I/\hbar < 1$ then we argue that the hopping rate from layer
to layer is

$$(\tau_\parallel I/\hbar)^2/\tau_\parallel,$$

which is proportional to $\tau_\parallel$, it being supposed that τ is caused by (inelastic)
collisions with phonons (or spin waves). We can see then that the T-dependence
of $\sigma_\parallel$ and $\sigma_\perp$, $d\sigma/dT$, should be similar. This appears to be the case for the
high-T_c superconductors well above T_c.

This conclusion is, we believe, valid only if $\tau_\parallel$ is caused by inelastic collisions
(corresponding to our L_i) and not if it results from scattering by impurities. In that
case, if $L_i > l$, the concept of a Fermi surface may not be valid simply because,
perpendicular to the planes, scattering is so strong that $\Delta k/k \sim 1$. But, unless
localization in the planes is significant, *activated* conduction across the planes is
not possible, because the Fermi energies in each plane will be the same. We
conclude that, however small the tunnelling factor, a conductivity less than
σ_{IR} ($\sim \frac{1}{3}e^2/\hbar a$, where a is the tunnelling distance) is unlikely.

In two dimensions, however, any disorder leads to localization, and in, for
instance, $La_{2-x}Sr_xCuO_4$ the random positions of the centres could lead to quite
strong localization. We have to ask, then, under what conditions can interaction

with the next planes delocalize the localized states. Suppose that ξ is the localization length in a given plane. This is associated with a time τ given by

$$1/\tau = D/\xi^2,$$

where D is the diffusion coefficient. The tunnelling time from one layer to the next is $\hbar/I$. We think then that delocalization can occur if

$$\hbar/I < \tau,$$

which gives

$$\hbar D/I\xi^2 < 1.$$

Since I is small, this cannot occur unless ξ is large; the disorder cannot have caused strong localization in the planes, doubtless because the charged centres are strongly screened.

10

Metal–Insulator Transitions in Liquid Systems

1 Introduction

In this book we have used the term "metal" to describe materials in which the conductivity tends to a finite value as the temperature T tends to zero, and "insulator" to describe those for which the resistivity tends to infinity. It therefore goes without saying that for a liquid no such sharp distinction can be made. Nevertheless, the study of conduction in liquids, and a comparison with amorphous metals, does in our view reveal points of considerable interest about metal–insulator transitions, which will be described in this chapter.

Our understanding of conduction in liquid metals is based on the theory put forward by Ziman (1961). This is a weak-scattering theory that has proved highly satisfactory for the description of normal metals for which $l \gg a$, and proves surprisingly satisfactory in many cases even when this is not the case (see Faber 1972, Mott and Davis 1979, Chap. 5). Most metal–insulator transitions in liquids occur in the regime where $l \sim a$. They can be induced in several ways.

(i) By expanding the metal at high temperatures, under pressure, a continuous transition from liquid to vapour can be observed. The work of Professor Hensel's laboratory at Marburg on mercury and caesium is reviewed in Section 4; it is maintained that a "Mott" transition occurs in caesium and a band-crossing transition in mercury. In the former case the disorder does not prevent the predicted discontinuity leading to a two-phase region, though it does affect the conductivity near the transition.

(ii) A change of composition can also lead to a metal–insulator transition. Solutions of alkali metals in ammonia are a typical example. Here a transition of Mott type (e.g. for one-electron centres) is thought to occur, though the phenomenon is complicated (as it is in caesium) by the tendency of one-electron centres to form pairs. The liquid system $Te_{1-x}Tl_{2+x}$ will also be discussed below.

229

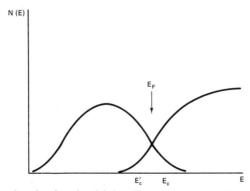

Fig. 10.1 Two overlapping bands with localized tails. A transition in a liquid can take place when the overlap is sufficiently large that states at E_F become delocalized, the mobility edges E'_c and E_c coinciding.

(iii) Other systems (e.g. In_2Te_3 and Ga_2Te_3) change their structure with increasing temperature, as do liquid Se–Te alloys and Te itself, where the chains break up, leading to metallic behaviour.

There is, we believe, a very important difference between liquid and amorphous metals, which was proposed by the present author (Mott 1985, 1989a). This is that there is evidence that quantum interference does not occur in liquids, so that the term $(k_F lg)^{-2}(1 - l/L_i)$ in equation (52) of Chapter 1 is absent. This is because, we argue, all collisions are inelastic, so that $l = L_i$. The experimental evidence and the theoretical arguments are set out in the next two sections. As a result, in principle, a minimum metallic conductivity exists, though the phenomenon is not sharp on account of the high temperature. The quantum interference effect is *not* absent in amorphous metals, there being ample evidence both for it and for the interaction effect of Altshuler and Aronov (Chapter 5, Section 6). This is described briefly in Section 7. In Section 4 we describe the behaviour of Hg and Cs at high temperatures and pressures, and in Section 5 that of metal–ammonia solutions.

Throughout we make use of the pseudogap model outlined in Chapter 1, Section 16. A valence and conduction band overlap, forming a pseudogap (Fig. 10.1). States in the gap can be Anderson-localized. A transition of pure Anderson type to a metallic state (i.e. without interaction terms) can occur when electron states become delocalized at E_F. If the bands are of Hubbard type, the transition can be discontinuous (a Mott transition).

A more complicated model has been used with success by Cutler and others, in which selenium and tellurium chains are broken with increasing temperature, giving rise to dangling bonds in equilibrium with valence alternation pairs; this is described in Section 6.

In liquid alloys of alkali metals with lead there are peaks in the plot of the resistivity ρ against composition—in KPb ρ rises to $\sim 1000\,\mu\Omega\,cm$

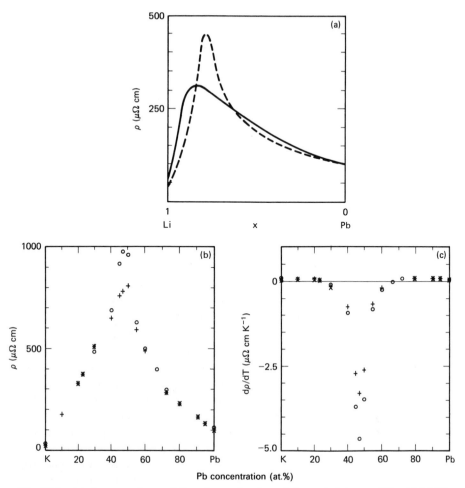

Fig. 10.2 (a) A comparison of the observed value of the resistivity ρ of liquid Li–Pb (dashed line) with the predictions of the Ziman theory (full line). From Pasturel and Morgan (1987). (b) Resistivity ρ and (c) $d\rho/dT$ of K–Pb alloys: ●, 400°C; ○, 375°C; +, 625°C. From Meijer *et al.* (1985).

($\sigma \sim 10^3\,\Omega^{-1}\,cm^{-1}$), much greater than σ_{IR}^{-1} if all valence electrons are free. Pasturel and Morgan (1987) account for this by quantum inference: Fig. 10.2(a) shows, for liquid Li–Pb, the resistivity calculated using the Ziman theory, compared with observed values. If, however, we are correct in believing that quantum interference does not exist in liquids then another explanation must be found. Meijer *et al.* (1985) (see also van der Lugt and Geertsma 1987), who carried out the measurements, proposed that for Li_4Pb, where the peak occurs, the four electrons from lithium are transferred to the lead, giving a nearly closed shell and consequently a low density of states. This is confirmed by the low value of the Knight shift. Their results for K–Pb are shown in Figs. 10.2(b, c). A discussion of the problem has also been given by Enderby and Barnes (1990).

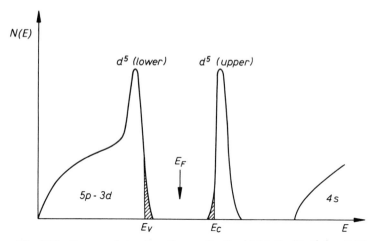

Fig. 10.3 Suggested density of states for liquid MnTe (Enderby 1990).

A very simple form of metal–insulator transition occurs in Mn_yTe_{1-y} at 1500 K (Barnes and Enderby 1985, Enderby and Barnes 1990). Figure 10.3 shows schematically the density of states when $y=0$. The Mn "upper Hubbard band" is narrow, but the lower one is hybridized with Se p-functions, and therefore conductivity is always p-type, facilitating interpretation of the thermopower. When $y=0$ the material is an insulator, but for any finite value of y there are $2y$ holes in the valence band. The behaviour is similar to that of $Mg_{3-x}Bi_{2+x}$ described in Chapter 7, Section 4. Figure 10.4 shows that the conductivity is of the form $ne^2\tau/m$, where the relaxation time τ does not vary much with the number of electrons. This suggests that inelastic collisions take place with atoms of the liquid, and that the frequency of such collisions is independent of n. The mean free path $l=v_F\tau=\hbar k_F\tau/m$, can thus be much less than the electron wavelength for inelastic collisions, so the Ioffe–Regel rule is not valid here (cf. Chapter 1, Section 6.3).

2 Absence of quantum interference in liquids; experimental evidence

Several pieces of experimental evidence can be cited. The first concerns the resistivity of the liquid alloys $Te_{1-x}Tl_x$, investigated by several authors (Cutler and Field 1968, Donally and Cutler 1972, Cutler 1977). We interpret their properties as follows. The Te^{2-} ion in the molecule $TeTl_2$ forms a closed shell. Thus wave functions near the bottom of the conduction band should be parabolic, and there will not be an extensive range of localized states for concentrations of thallium above that in $TeTl_2$ in the conduction band. A degenerate gas of non-localized electrons is expected. The conductivities confirm this, being in the metallic range and little dependent on T, and the thermopower is negative and small. It shows a discontinuous change to a positive value when the

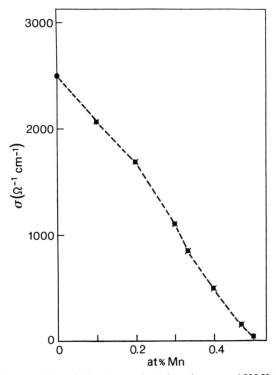

Fig. 10.4   Conductivity of Mn_yTe_{1-y} plotted against y at 1500 K (Enderby 1990).

Tl concentration falls below this value, and the resistivity jumps to about $10^{-2}\,\Omega$ cm. We have here a degenerate gas of holes, moving in a highly disordered medium, and with a conductivity well below σ_{IR} (the Ioffe–Regel value) and approaching σ_{min}. In Fig. 10.5 we reproduce a curve from Cutler (1977), plotting the Pauli paramagnetic susceptibility against the square root of the conductivity, $\sigma^{1/2}$, for specimens of different compositions and temperatures. The former is proportional to $N(E_F)$, and thus, in the pseudogap model, to our constant g. We see that σ is proportional to g^2. Thus the interference term in equation (52) of Chapter 1 appears to be absent.

The second piece of evidence comes from the observations of Warren (1970a, b, 1971, 1972a, b) on the Knight shift in In_2Te_3 and Ga_2Te_3. These materials are among those that make a gradual transition to the metallic state as the temperature is raised. In the case of liquid Te, as the temperature rises, the change is known to be accompanied by an increase in coordination number from 2 to 3 (Cabane and Friedel 1971). In Warren's work g is deduced directly from the Knight shift K, which is proportional to g, with g and K increasing with T. Figure 10.6 is a plot of electrical conductivity versus g (with varying temperature). There is a rather abrupt change of slope round $g \approx 0.2$, which Warren interprets as the onset of localization. Warren deduces for the "minimum metallic conductivity"

$$\sigma_{min} = 200 \pm 70\,\Omega^{-1}\,cm^{-1},$$

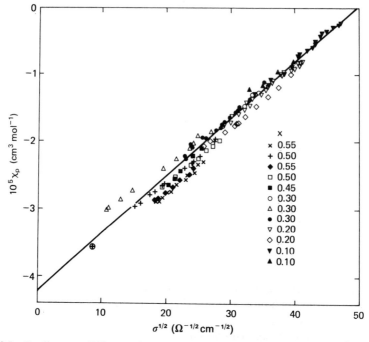

Fig. 10.5   Pauli susceptibility χ_P of $Te_{1-2}Tl_x$ liquid alloys plotted against $\sigma^{1/2}$ for various temperatures and compositions (Cutler 1977).

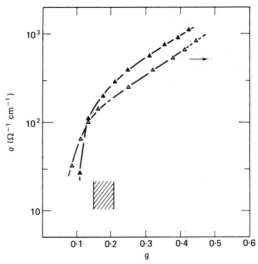

Fig. 10.6   Electrical conductivity plotted against the factor g deduced from the Knight shift for In_2Te_3 (▲) and Ga_2Te_3 (△) (Warren 1971). Localization starts in the shaded region.

in good agreement with our predicted value. In the region before localization the relation $\sigma \sim K^2$ is well satisfied, as also is the prediction that the Korringa broadening is proportional to σ and thus to K^2.

Another material in which there is proportionality between σ and g^2, the latter deduced from the Knight shift, is liquid tellurium. The conductivity between 1300 and $2700 \, \Omega^{-1} \, cm^{-1}$ is less than the Ioffe–Regel value for six electrons per atom $(6000 \, \Omega^{-1} \, cm^{-1})$, which we interpret in terms of a pseudogap that gradually fills in with increasing temperature as the coordination number increases. Figures 10.7(a, b) show that $\sigma \sim K^2$, the data coming from Warren (1971).

Kao and Cutler (1988) found in $Se_x Te_{1-x}$ between $x = 0.1$ and 0.2, where the metallic approximation is valid, that σ is proportional to E_F and thus, if $N(E) \propto E^{1/2}$, to $[N(E_F)]^2$, in agreement with our hypothesis for the materials discussed above (provided that the "tail" to $N(E)$ does not affect $N(E)$ for these values of x).

Further evidence is given in Section 5.5 for solutions of alkali metals in ammonia.

A striking difference between the behaviour of the Knight shift at an Anderson transition in liquids and in solids is expected for the pseudogap model (Mott 1972b, 1974c). As the wave functions at the Fermi energy change from extended to localized owing to the variation of any parameter, no discontinuity in $N(E_F)$ or the Pauli paramagnetism is expected (see Thouless 1972). On the non-metallic side of the transition, then, the Pauli susceptibility is $2\mu^2 N(E_F)$, as it is on the metallic side. But if, under the influence of a magnetic field H, $H\mu N(E)$ electrons change their spin direction per unit volume, these will occupy a fraction of the volume equal to

$$\tfrac{4}{3}\pi H \mu N(E) \alpha^{-3}. \tag{1}$$

So, as regards the Knight shift, only this fraction of the nuclei will feel the resultant field. Those that do will be acted on by a strong field that is independent of H but increase as α increases. Thus the NMR line should show a splitting ΔH independent of H, but increasing rapidly as $c_0 - c$ increases. The two lines will be of comparable intensity when (1) approaches unity. Since $2N(E)\mu^2$ is the Pauli susceptibility χ, this occurs when

$$H\chi \sim \mu \alpha^3. \tag{2}$$

If α^{-1} is greater than the distance between centres, as it must be near the transition, then this should occur for fields much smaller than those required to saturate the moments. The expected behaviour is shown in Fig. 10.8.

In liquids, if E_F crosses the mobility edge E_c, no discontinuity in the Knight shift is expected, because the timescale of the nuclear resonance is long compared with the time in which atomic movement will change the positions of the localized states. If Warren's interpretation of his measurements on liquid tellurium alloys is accepted, there is certainly no discontinuity in K when this happens.

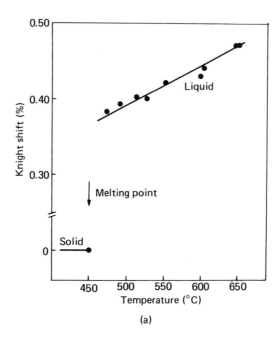

(a)

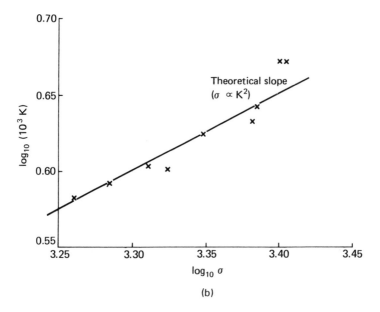

(b)

Fig. 10.7   (a) Knight shift K of liquid tellurium (Cabane and Froidevaux 1969). (b) Plot of $\log_{10}(10^3 K)$ versus $\log_{10} \sigma$, showing the linear relationship $\sigma \propto K^2$, where K is the Knight shift.

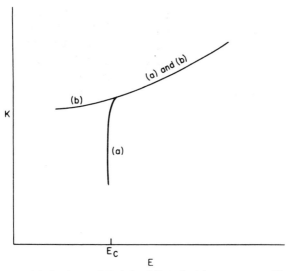

Fig. 10.8 Expected behaviour of Knight shift K in (a) a non-crystalline solid and (b) a liquid when E_F crosses E_c, giving an Anderson transition.

3 Absence of quantum interference; theory

The experimental evidence, then, suggests that quantum interference is absent in liquid metals. At first sight this might seem to contradict Ziman's (1961) theory of liquid metals, in which waves scattered by different atoms do interfere; this, however, is not so. Following arguments of Baym (1964), Greene and Kohn (1965) and Faber (1972), one should not use in that theory the Fourier transform $S(k)$ of the instantaneous pair-distribution function, but rather

$$\int S(k, \omega) \frac{\beta \omega \, d\omega}{1 - e^{-\beta \omega}},$$

where $\beta = (k_B T)^{-1}$. The T-dependent term makes only a small difference (Greene and Kohn 1965), and it can be shown that

$$S(k) = \int S(k, \omega) \, d\omega \tag{3}$$

is the Fourier transform of the pair distribution for the instantaneous positions of the atoms. Here $S(k, \omega)$ is the frequency-dependent transform. It is shown in textbooks on neutron diffraction (e.g. Squires 1978, p. 31) that one can represent the scattering after the excitation of $1, 2, \ldots$ phonons, and then of course each term allows interference between scattered waves as in the Ziman theory; on adding all these scattered waves, one arrives at (3), as long as the phonons are not localized. In the multiple-scattering problem from which one deduces the absence of quantum interference in liquids, one considers the two inelastic collision processes by the whole liquid, and the resulting waves cannot interfere.

An alternative explanation is given by Mott and Kaveh (1990). The work of Afonin *et al.* (1987) has already been mentioned, showing that quantum interference is not incompatible with inelastic scattering if the energy loss is small. Mott and Kaveh (1990), in a paper appearing too late for full discussion here, show that to explain its absence in liquids one has to suppose that the path shown in Fig. 1.23 includes many collisions.

4 Fluid mercury and caesium at high temperatures

In the next two sections we review experimental work on the conductivity and magnetic properties of the following materials:

(a) mercury at high temperatures and expanded volume, in which a transition to metallic properties occurs as a result of the volume changes; with two electrons per atom, mercury is thought to undergo a band-crossing transition;

(b) similar work on caesium, which with one electron per atom is thought to undergo a Mott transition;

(c) solutions of alkali metals in ammonia; the metal ion is solvated by the ammonia molecules, and the electron is trapped in a cavity (see e.g. Thompson 1976, 1985); here also we expect a Mott transition.

We discuss these materials under the following assumptions.

In these materials, in spite of the disorder, a band-crossing or Mott transition can take place much as in crystals. Thus a large discontinuity in σ at a metal–insulator transition can be expected in caesium and $NaNH_3$, but a much smaller one in mercury (Chapter 4, Section 3). The two-phase region with a critical point in $NaNH_3$ is well known (Fig. 10.12), and at the critical point in this material and in caesium the conductivity σ is of order $3 \times 10^2 \, \Omega^{-1} \, cm^{-1}$, and is close to σ_{min}. We suppose that, for both, the critical points are a consequence of the transition, as illustrated in Fig. 4.3. For mercury, on the other hand, the conductivity at the critical point ($\sim 10^{-3} \, \Omega^{-1} \, cm^{-1}$; Hensel 1971) is five orders of magnitude smaller; the critical point here, which results from the overlap of the $6p^2$ outer shells on adjacent atoms, is of the same nature as in argon. A discussion of the two kinds of critical point was given first by Landau and Zeldovich (1943), with the suggestion that *two* critical points might be observed.

In fluids, we have already postulated that quantum interference does not occur. Of course, in liquids no sharp metal–insulator transition can be expected, and therefore we *define* the transition as occurring when $\sigma = \sigma_{min}$, the quantity given by equations (50 a, b) of Chapter 1. We therefore believe that it is legitimate to write for the thermopower

$$S = \frac{k_B}{e} \left(\frac{E}{k_B T} + A \right), \tag{4}$$

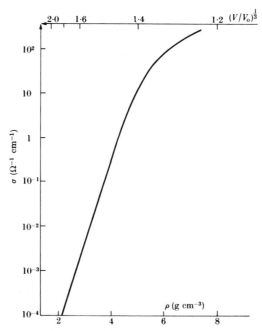

Fig. 10.9 Specific conductivity of mercury at 1550°C as a function of density ρ (Hensel and Frank 1968). V is the volume and V_0 that at normal temperature and pressure.

with $A = 1$ (Cutler and Mott 1969, Mott and Davis 1979), and suppose (unless polarons are formed) that

$$\sigma = \sigma_0 e^{-E/k_B T}, \quad \sigma_0 \approx \sigma_{\min}, \tag{5}$$

with the same activation energy. Of course, E may change with temperature, on account of a change of structure, but we do not believe such effects to be very important; they will not affect the equality of the two activation energies.

To determine σ_0, then, we may eliminate E from (4) and (5). Taking $|S|e/k_B = 1$ at the transition and $A = 1$, σ should be the value of σ when $S = 83\,\mu\text{V K}^{-1}$:

$$\log_{10} \sigma = \log_{10} \sigma_0 - \left(\frac{|S|e}{k_B} - A \right) \Big/ 2.3. \tag{6}$$

We expect to find σ_0 of the order $200\,\Omega^{-1}\,\text{cm}^{-1}$ for Hg and Cs, and somewhat smaller for NaNH$_3$.

Turning now to fluid mercury and caesium, a fairly recent review is given by Freyland and Hensel (1985). Figure 10.9 shows the classic results of Hensel and Franck (1968) on the conductivity of mercury as a function of volume V; the metal–insulator transition occurs when $V/V_0 \approx 1.3$. Schönherr et al. (1979) measured the conductivity σ and thermopower S of mercury for conductivities between 200 and $5\,\Omega^{-1}\,\text{cm}^{-1}$ for mercury. Their results for σ as a function of S are shown in Fig. 10.10(a). The slope is exactly as predicted, and σ_0 is in the range

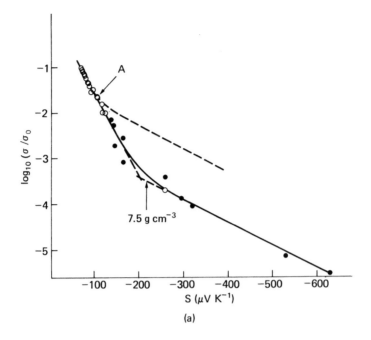

(a)

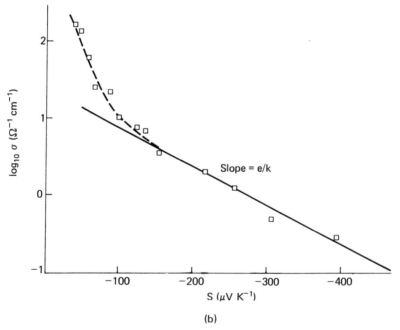

(b)

Fig. 10.10   (a) Logarithm of conductivity σ/σ_0, with $\sigma_0 = 10^4\,\Omega^{-1}\,cm^{-1}$, of expanded fluid mercury (Schönherr *et al.* 1979) and (b) logarithm of conductivity σ of liquid selenium, using data of Perron (1967), plotted versus thermopower S.

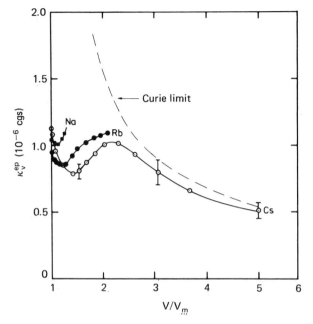

Fig. 10.11 Volume spin susceptibility of expanded alkali metals versus reduced volume (V_m = volume at melting point). From Freyland and Hensel (1985).

140–200 Ω^{-1} cm^{-1}. Freyland *et al.* (1974) obtained similar results for caesium, with $\sigma_0 \sim 300\,\Omega^{-1}$ cm^{-1}. All of these results are in fair agreement with our predictions.

For selenium the value is smaller, about $20\,\Omega^{-1}$ cm^{-1} (see Fig. 10.10(b) and also Kao and Cutler 1988). In Section 6 we suggest that this is a consequence of the chain structure of this material.

The discussion given by Freyland and Hensel (1985), citing band calculations for Cs by Warren and Mattheiss (1984), shows that in the metallic state near the transition in this material there is strong evidence for the enhancement (Chapter 4, Section 6) of the magnetic susceptibility predicted by Brinkman and Rice (1970b). Figure 10.11 shows the high-temperature spin susceptibility for alkali metals, plotted against reduced volume. The conductivity reaches $500\,\Omega^{-1}$ cm^{-1} (σ_{min}) for $V/V_m \sim 3$. So the peak in κ is explained as Brinkman–Rice enhancement.

The effect is not large. Since σ_{min} varies as a^{-1}, and a should be the distance between the effective current carriers, only a small enhancement of κ is expected.

The proximity between the curves for the Curie limit and that observed show that the proportion of pairs (Cs$_2$ molecules) is not large.

Finally, we mention that there is some doubt whether the transition in mercury is of normal Anderson type; Turkevich and Cohen (1984) have proposed a model in which an excitonic insulator is formed, for conductivities below $\sim 10^2\,\Omega^{-1}$ cm^{-1}.

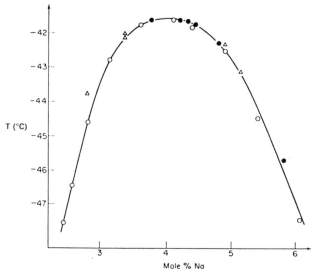

Fig. 10.12   Coexistence curve for Na–NH$_3$ (Chieux and Sienko 1970).

5   Metal–ammonia solutions

5.1   Introduction

Alkali metals can be dissolved freely in liquid ammonia without chemical reaction. The resulting solutions have been regarded both as scientific curiosities and as subjects for serious investigation since their discovery by Weyl (1864). Concentrated solutions in the range 7–20 MPM have a metallic bronze colour and the high electrical conductivity characteristic of a liquid metal; dilute solutions have a bright blue colour and for concentrations below 10^{-3} MPM behave as ideal electrolytes with solvated metal ions and solvated electrons as the ionic species. Somewhere between, a metal–insulator transition takes place; this section gives a discussion of this transition. We compare what is happening with the situation in fluid caesium.

As we shall see below, for dilute solutions the electron is not attached to the alkali ion but is trapped in a cavity, around which the ammonia is polarized. The problem of the metal–insulator transition, then, is one of a random array of one-electron centres, as in a doped single-valley semiconductor. On the other hand, the disorder is less because the strong overlap between the wave functions of some pairs of centres characteristic of doped semiconductors is absent. In doped semiconductors there is no discontinuity in ε_2 at the transition. As explained in Chapter 5, this may be because of the very strong disorder or, in many-valley systems, because of self-compensation. In metal–ammonia solutions, as in the fluid alkali metals discussed in Section 4, both are absent.

The most striking feature of M–NH$_3$ is the immiscibility of concentrated and dilute solutions; a typical phase-separation curve is shown in Fig. 10.12 (Chieux and Sienko 1970). This immediately suggests that in M–NH$_3$ there *is* a discontinuity in *n* (or rather there would be at zero temperature), leading to a free energy-composition curve with a kink as in Fig. 4.2. This was first suggested by Mott (1961).

Metal–ammonia solutions are to be compared with molten salt solutions of the type KI–K, which show a two-phase region (Bredig 1964, Warren 1985), and with caesium vapour. This was emphasized particularly by Krumhansl (1965).

In metal–ammonia solutions most experimental work has been carried out at temperatures above the consolute point and here we believe that the system can be treated similarly to a Mott transition in a solid. However, there are differences.

(i) Near the consolute point, long-range fluctuations in density play a role.

(ii) There is evidence that the solvated electrons form pairs, presumably in association with a cation. The evidence for this is that the Pauli susceptibility and Knight shift drop as we approach the transition from the metallic side, and on the non-metallic side the Curie susceptibility is smaller than it would be if all moments were free, except at low concentrations. This makes the transition more like a band-crossing transition than one in which Hubbard bands cross, for which the susceptibility would *increase*, the bands being formed from molecular orbitals of odd and even parity on the pairs (like bands in liquid H$_2$). This in no way alters our conclusion that the two-phase region results from a metal–insulator transition with a discontinuity in the number of current carriers, so long as the oscillator strength between the two bands is small, so that the discontinuity in *n* is large. We believe that the separation between the bands is mainly determined by the Hubbard *U*, and that the situation is similar to that described for VO$_2$ in Chapter 6, Section 5.

We turn now to a more detailed discussion of the metal–ammonia system.

5.2 Very dilute solutions

Liquid ammonia becomes conducting on dissolving small amounts of alkali or alkaline-earth metal. The dissolution is reversible; no chemical reaction takes place. It follows immediately that the metal atoms dissociate into positive ions and electrons. The nature of these solvated electrons is discussed in this section.

Upon dissolving the metal, a broad optical absorption line appears, peaked at 0.85 eV, and with a tail extending into the visible, which gives the characteristic blue colour (Fig. 10.13). The absorption does *not* depend on the nature of the solute, showing that the solvated electron dissociated from the cations is responsible for the absorption. The absorption spectrum is almost independent of concentration up to $\sim 10^{-1}$ MPM.

The solvated electron is responsible for a single, extremely narrow, structureless spin resonance line in the dilute solutions with a *g* value of 2.0012. Integration of its intensity gives a static spin susceptibility tending towards the

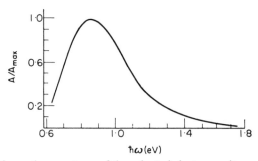

Fig. 10.13 Absorption spectrum of the solvated electron as it appears in dilute Na–NH$_3$ solutions (Burow and Lagowski 1965). The high-energy tail is the source of the characteristic blue colour of dilute solutions. From Cohen and Thompson (1968).

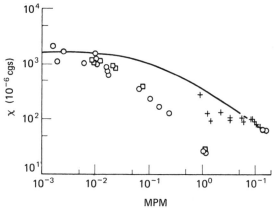

Fig. 10.14 Measured static susceptibilities for Na and K solutions in NH$_3$ and the calculated spin susceptibility of a set of independent electrons at 240 K (solid line). The diamagnetic contribution of the NH$_3$ molecules has been eliminated from the measured total susceptibility by using the Wiedemann rule. Since the total susceptibility is quite small in concentrated solutions, the errors may be large. ○ represent data of Huster (1938) on Na solutions at 238 K; □ represent K–NH$_3$ data of Freed and Sugarman (1943) at the same temperature; and + represent data of Suchannek et al. (1967) at room temperature for Na–NH$_3$ solutions. From Cohen and Thompson (1968).

free-spin value at the lowest concentration studied. However, for concentrations above 10^{-3} MPM there is a marked decrease in the spin susceptibility per electron (Fig. 10.14). This suggests that between 10^{-3} and 10^{-1} MPM the solvated electrons form some sort of association such that spins are paired, leading to diamagnetism, but which at the same time does not affect the optical absorption. The nature of this association, already mentioned in Chapter 5, Section 1, has long been a matter of dispute and will be discussed in Section 5.3.

In very dilute solutions the diffusion coefficient for solvated electrons at room temperature is

$$D(e^-) = 2.66 \times 10^{-4} \, \text{cm}^2 \, \text{s}^{-1}$$

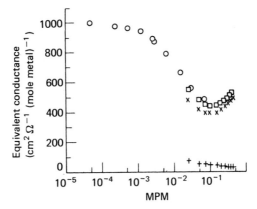

Fig. 10.15 Conductance per ion pair of metal–ammonia solutions. The ratio of electrical conductivity to the concentration of metal (equivalent conductance) is shown as a function of concentration. ○ represent data of Kraus (1921) and ☐ data of Dye *et al.* (1960), both at 240 K and in Na–NH$_3$ solutions. + and × represent the equivalent conductances assigned to positive and negative carriers respectively by Dye on the basis of transference-number measurements. From Cohen and Thompson (1968).

(Catterall 1970), which is considerably higher than that for the solvated cation

$$D(Na^+) = 4.66 \times 10^{-5}\, cm^2\, s^{-1}.$$

The mechanism is described below. The mobilities μ and conductivities for concentrations below 10^{-3} are related to D by the Einstein relation

$$\mu = eD/k_B T.$$

For higher concentrations the equivalent conductance (conductance per ion pair) falls off, as shown in Fig. 10.15. This has been interpreted as suggesting association into neutral complexes, though we shall put forward a different possibility.

Finally, a striking property of metal–ammonia solutions is the large expansion of the liquid due to the solvated electrons. The apparent volume of the solvated electron remains roughly constant up to the metallic range, then shows a slight increase. It is about $100\, cm^3\, mol^{-1}$. It is this effect that has led to the hypothesis that the electron forms a cavity for itself; a cavity of radius 3.2 Å accounts quantitatively for the excess volume. A model in which the electron moves in a cavity, and the surrounding liquid is polarized or solvated as it is round a cation, was first put forward by Jortner (1959), who showed that it was able to account for the absorption spectrum. Jortner's model, as modified by Mott (1967), Cohen and Thompson (1968) and Catterall and Mott (1969), will now be described.

Suppose that the electron forms for itself a cavity of radius R. Let the energy of an electron at the bottom of the conduction band of NH$_3$ be V_0. Then at the

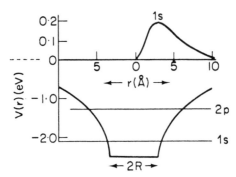

Fig. 10.16 Potential seen by a solvated electron according to the model of Jortner (1959). The wave function $r\psi$ of the electron is also shown. The optical absorption is due to the 1s–2p transition . The radius R of the cavity is approximately 3.2 Å. From Cohen and Thompson (1968).

boundary of the cavity there should be a potential step V_0. For $r > R$ the electron will see a potential

$$V(r) = -\frac{e^2}{r}\left(\frac{1}{\kappa_\infty} - \frac{1}{\kappa}\right),$$

giving a total potential as shown in Fig. 10.16. Jortner showed that this model was able to account for the absorption line shown in Fig. 10.13 as being due to 1s→2p transitions by the electron; the 2p-level will be greatly broadened by thermal oscillation of the positions and orientations of the NH_3 dipoles. Jortner obtained agreement with experiment by taking $V_0 \approx 0$; in other words, the energy of the bottom of the conduction band in NH_3 is about that of an electron in free space. Cohen and Thompson (1968) showed that this is likely on a priori grounds.

Blandamer et al. (1964) pointed out that the absorption spectrum of iodine ions (I^-) in NH_3 has its maximum at $h\nu = 4.0\,\text{eV}$; the difference from that for an electron in a cavity, $4.0–0.8 = 3.2\,\text{eV}$, corresponds well to the electron affinity of iodine. In water the maxima for both I^- and a solvated electron are shifted by $0.8\,\text{eV}$ to higher frequencies; we deduce that the energy of the bottom of the conduction band in water is about $0.8\,\text{eV}$.

As regards the origin of the cavity, several explanations are possible in principle. If V_0 is positive then the energy will be lowered by about V_0 if a cavity is formed. This is the origin of the cavities formed in liquid helium. This can only be the correct mechanism if V_0 is greater than the surface energy $4\pi\gamma a^2$, where a is the radius and γ the surface energy per unit area. With $a = 4\,\text{Å}$ and $\gamma = 32\,\text{erg}\,\text{cm}^{-2}$, this gives $V_0 \approx 0.35\,\text{eV}$. But this is certainly an underestimate since not all of $\int \psi^2\,d^3x$ is within the cavity, and values of V_0 of order $1\,\text{eV}$ would probably be necessary. These are unlikely, and an alternative explanation introduced by Catterall and Mott (1969) is based on the concept of the Bjerrum defect. A Bjerrum defect (Bjerrum 1951) is a site in a crystal such as ice at which the positive charges in two dipole molecules point towards each other. The energy of such a defect in ice is estimated to be about 0.34 eV. Around a self-trapped electron the direction of polarization must change rapidly, and it is supposed that energy will be lowered if a cavity is formed on account of the repulsion between like charges on NH_3 dipoles.

Quite a different model has been used, taking a molecular rather than a continuum point of view, and may give results nearer to the truth. In this analysis there is no actual cavity, but rather a volume of low density (McAloon and Webster 1969, Howat and Webster 1972).

Since the publication of the first edition of this book, there has been extensive theoretical work on the nature of the cavity (Kevan and Webster 1976, Thompson 1984, 1985, Brodsky and Tsarevsky 1984). On the whole, the model is not greatly changed.

As already stated, the diffusion coefficient and mobility of a solvated electron, though much smaller than that of a free electron in a conduction band, are some 4–5 times larger than for the solvated cation. There is some doubt whether the electron carries along a solvated shell with it, and there has been discussion in the literature of mechanisms by which molecules drift across the cavity, changing their orientation as they do so.

5.3 The dilute region (10^{-3}–10^{-1} MPM)

The salient features observed in this range of concentration are as follows.

(i) There is little change in the optical absorption coefficient.

(ii) The conductivity per electron drops (Fig. 10.15) to about 20% of the value for very dilute solutions. However, the mobility of the cation does not seem to be affected.

(iii) There is a marked drop in the spin susceptibility, either deduced from ESR measurements or obtained directly. This is particularly marked at low temperatures (Fig. 10.14). It appears that at the higher concentration only about 20% of the cavities carry free spins.

The small change in the optical absorption and the low mobility suggest strongly that we are dealing here with solvated electrons with properties little different from the very dilute case. There has been much discussion in the literature about whether the drop in the conductivity and in the number of free spins ought to be explained in terms of a theory similar to that of Debye and Hückel for strong electrolytes or whether diamagnetic complexes are formed. While the problem is not fully resolved, we think that the evidence strongly suggests that a pair of cavities can combine weakly in association with a cation to form a "molecule", the attraction being due to homopolar forces, as in the hydrogen molecule H_2. We shall call such a "molecule" the "molecular dimer", to be distinguished from the "atomic dimer" formed when two electrons are in the same cavity. The properties of the "molecular dimer" should be as follows:

(a) it carries an electronic charge $2e$, though perhaps partly neutralized by a cation;

(b) it is diamagnetic;

(c) its binding energy should not be high, perhaps in the range 0.1–0.2 eV, so that some dissociation could occur, nearly completely at low concentrations.

Such a model can clearly account for the drop in the magnetic susceptibility. The drop in the conductivity is to be expected because the molecular dimer should be considerably less mobile than the monomer, the solvation should be stronger, and it will probably be more difficult for the NH_3 molecule to drift across the pair of cavities. The alternative explanation is of course the formation of neutral complexes with the cations; but if the cations retain their original mobility, this seems ruled out.

A review by Schindewolf (1984) supports the idea that a complex forms with configuration $e^- M^+ e^-$, where M^+ is the solvated metal ion.

Mott (1980 a) has given a further argument against the assumption that two electrons in the same cavity form the lowest state of the system. If this were so then the transition would be of band-crossing type, and one would not expect the observed two-phase region.

5.4   The metallic region (>7 MPM)

For the moment we pass over the intermediate range and consider highly concentrated solutions. Figure 10.17 shows the conductivity over the whole range, and for concentrations of about 20 MPM this reaches the value for liquid mercury ($10^4 \Omega^{-1} cm^{-1}$), though the concentration of electrons is much less. There is no doubt that these are normal liquid metals; the quantity $k_F l$ reaches the value 30. Here l is the mean free path deduced from the expression (equation (29) of Chapter 1)

$$\sigma = S_F e^2 l / 12\pi^3 \hbar$$

and from the observed conductivity. According to the analysis of Ashcroft and Russakoff (1970), the resistivity is due to scattering by solvated cations and by the dipole moments of ammonia molecules not bound to cations; the drop in the number of the latter for high concentrations of metal accounts for the rapid increase in the conductivity. Catterall and Mott (1969) gave a qualitative account of the temperature dependence of the conductivity, which is positive.

Cohen and Thompson (1968) considered that electron–electron (Landau–Baber) scattering (see Chapter 2, Section 6) is also important, because the Fermi energy is small compared with that of a normal metal. In the liquid, because the

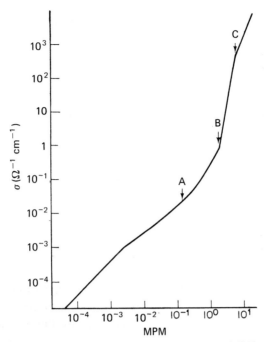

Fig. 10.17 Electrical conductivity of solutions Na in NH_3 at 240 K, due to Kraus (1921). "A" marks the minimum in conductance per ion pair shown in Fig. 10.15. From Cohen and Thompson (1968).

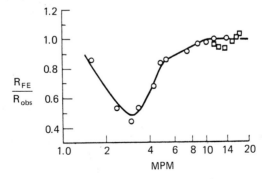

Fig. 10.18 Hall coefficient R for metal–ammonia solutions, plotted as a function of concentration in the form R_{FE}/R_{obs}. From Cohen and Thompson (1968).

Fermi surface is spherical, this does not give rise to a term in the resistance proportional to T^2, but it does in the solid $Li(NH_3)_4$ (McDonald and Thompson 1966, Mott and Zinamon 1970). However, in the liquid it does affect the heat conductivity, and indeed there is an anomaly in the Wiedermann–Franz ratio that Cohen and Thompson explain in this way.

The Hall coefficient is normal in the metallic region, but drops off as we approach the metal–non-metal transition (Fig. 10.18). It is found to be temperature-independent.

A marked effect is that the large increase in volume with increasing concentration of metal continues for the high concentrations described here. Cohen and Thompson explained this by supposing that cavities are still formed, the metallic electrons having large values of ψ in the cavities, so that the density of states is as in an impurity band. They also supposed that the positive value of $d\sigma/dT$ is due to a smearing out of the impurity band as the temperature is raised. While this is possible near the transition, we think it unlikely for high concentrations, where the mean free path is so large that no tight-binding band due to cavities can exist. An alternative qualitative description of the expansion in terms of Bjerrum defects was given by Catterall and Mott (1969), who supposed that at high concentrations the rate of change of the polarization of NH_3 between *cations* is so great that an expansion of Bjerrum type must occur.

5.5   The intermediate region (10^{-1}–5 MPM) and the metal–insulator transition

The transition between the metallic and non-metallic regions is, as we have stated, in many ways similar to that in doped semiconductors, but there are certain differences, which we believe are due to the pairing of cavities to form what we have called "molecular dimers". This intermediate region has two parts.

(i)   From (say) 6 to 3 MPM the Hall coefficient rises above the free-electron value (Nasby and Thompson 1970) and the conductivity drops to a value in the range 10–$100\,\Omega^{-1}\,cm^{-1}$.

(ii)   From 3 to (say) 0.3 MPM the conductivity is not in the metallic range, but we believe that the material behaves like an intrinsic semiconductor, current being carried by electrons in the upper band. This, as we have seen, is not a Hubbard band, but the band formed from molecular orbitals for an extra electron on molecular dimers. In this range $d \ln \sigma/dT$ shows a strong maximum at about 1 MPM (Fig. 10.19), and there is also a minimum in $d \ln \sigma/dp$; in addition, there is a change in sign of dS/dT, where S is the thermopower.

In the first range Acrivos and Mott (1971) and Acrivos (1972) supposed that the material is metallic, as the observed conductivities suggest, that the two bands overlap and that the factor g defined by

$$g = N(E_F)/N(E_F)_{free}$$

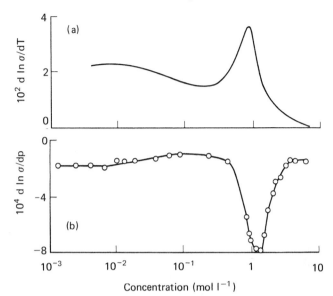

Fig. 10.19 Conductivity in metal–ammonia solutions (Schindewolf *et al.* 1966): (a) $10^2 \, d \ln \sigma/dT$; (b) $10^4 \, d \ln \sigma/dp$. From Catterall and Mott (1969).

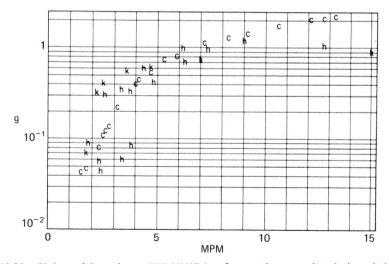

Fig. 10.20 Values of the ratio $g = N(E_F)/N(E_F)_{free}$ for metal–ammonia solutions deduced from Knight shift (k), conductivity (c) and Hall constant (h). (Data provided by Dr J. V. Acrivos; Hall data from Nasby and Thompson (1970) and Vanderhoff and Thompson (1971).)

is less than unity. They supposed that in this range

$$\sigma \sim g^2, \quad \chi_{el} \propto g, \quad K \propto g,$$

$$R_H = \frac{C}{necg}, \quad C \approx 0.7.$$

χ_{el} is the electronic contribution to the paramagnetism, obtainable either from the ESR signal or by direct measurements with subtraction of the diamagnetic term; R_H is the Hall coefficient, the formula being due to Friedman (1971). K is the Knight shift. g is found to drop from 1 to about 0.3 in this range and σ is then about $50\,\Omega^{-1}\,cm^{-1}$. This then should mark the metal–insulator transition. A plot of g deduced from the various data is shown in Fig. 10.20. This we consider further evidence for the absence of quantum interference in liquids (Section 2).

A striking fact about these results is that χ_{el} *decreases* with decreasing concentration; in doped semiconductors it increases as discussed in Chapter 5, apparently owing to the presence of free moments. This is strong evidence that our "molecular dimers" form and that they are diamagnetic, so that the metal–insulator transition is of the band-crossing type discussed in Chapter 1, Section 6. One envisages two bands: one for an extra electron on a molecular dimer, and one for a hole in a molecular dimer. The metal–insulator transition should occur when these overlap, though for "frozen" solutions one would expect variable-range hopping due to Anderson localization when the overlap is small, though this is unlikely to be observable in liquids.

Apart from the drop in the susceptibility as we approach the transition, the increase in the Hall-coefficient ratio (Fig. 10.18) is further evidence for the model; as we have seen, evidence from doped semiconductors shows that R_H does not deviate from the free-electron form when E_F lies in the pseudogap for current carriers in a highly correlated gas (unless antiferromagnetic order sets in). Also, the susceptibility *increases* with temperature; this may be because the molecular dimers dissociate.

We turn now to the range below 3 MPM, where, as shown in Fig. 10.19, the conductivity rather suddenly becomes sensitive to temperature and pressure. Also, in this region the thermopower S increases rapidly from its "metallic" value ($10-12\,\mu V\,K^{-1}$), which increases with T, to a value of about $80\,\mu V\,K^{-1}$, which *decreases* with T (Damay et al. 1969). At this point, in our view, the metal–insulator transition occurs. Before $d\sigma/dT$ rises, current is carried by electrons at the Fermi energy, the two bands due to holes and extra electrons on the dimers overlapping. When states become Anderson-localized, we do not think that hopping sets in; the excitation energy to the mobility edge is low enough at the temperature concerned for current to be carried mainly by electrons at the mobility edge. Hopping conduction is therefore not observed, and there is a sharp transition from truly metallic to truly semiconducting behaviour. Perhaps the clearest evidence for this is the change of sign of dS/dT.

The fact that the pressure coefficient of conductivity is large and negative in this region is unexpected; pressure will normally decrease a band gap. However,

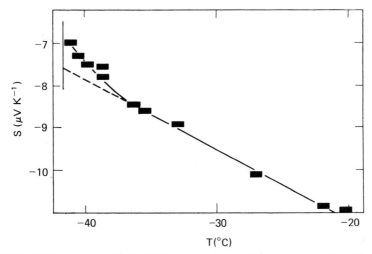

Fig. 10.21 Thermopower of Na–NH$_3$ as a function of temperature T at the critical concentration, showing deviation from the metallic formula as T approaches the consolute point (Damay 1973).

in spite of the formation of molecular dimers, the gap (as in VO$_2$) will be mainly due to the Hubbard U, and pressure will doubtless make the cavities smaller and increase U.

Below about 0.3 MPM the current is carried by the mass movement of the solvated electrons; the range in which semiconducting behaviour occurs is thus quite small.

5.6   Effects of the two-phase region

We note that at the consolute point the conductivity is still metallic, the appearance of an activation energy ϵ_2 occurring for somewhat lower concentrations. The reason for this, in our view, is as follows. The consolute point should occur approximately at the same concentration as the kink in the free-energy curve of Fig. 4.2, namely that at which the concentration n of carriers is of order given by $n^{1/3}a_H \approx 0.2$. Above the consolute point there is no sudden disappearance of the electron gas as the concentration decreases; its entropy stabilizes it, so metallic behaviour extends to lower concentrations, until Anderson localization sets in. Conduction, then, is due to excited electrons at the mobility edge, as discussed above.

At the time of writing, the only evidence for critical fluctuations near the consolute point known to us comes from the work of Damay (1973). The thermopower of Na–NH$_3$ plotted against T at the critical concentration is shown in Fig. 10.21. We conjecture that this behaviour is due to long-range fluctuations between two *metallic* concentrations, and that near the critical point, where the fluctuations are wide enough to allow the use of classical percolation theory, the more metallic regions determine the thermopower, which therefore falls. The results of Fig. 10.21 show rather clearly that the effect of critical behaviour on electrical properties extends only a few degrees above the consolute point.

Metal–ammonia solutions show a dramatic increase in the dielectric constant as the concentration approaches that for a metal–insulator transition from below. We think this has to be interpreted as in Chapter 5, Section 8. Investigations of this behaviour through ESR measurements in Li_xNH_3 have been made by Damay et al. (1988), with a theoretical interpretation given by Leclercq and Damay (1988).

There have been many theories of metal–ammonia solutions that differ from that presented here. Cohen and Thomson (1968) and more recently Cohen and Jortner (1973) supposed that large fluctuations in concentration occur over large distances, sufficient to invalidate the pseudogap model and substitute a semiclassical percolation theory. For the reasons given above, we think that this is only likely to be so within a few degrees of the critical point.

As pointed out by Thompson (1985), the calculated value of σ_{min} at the consolute point is about $70\,\Omega^{-1}\,cm^{-1}$. This is very close to the observed value of the conductivity at this point, which is changing rapidly with concentration.

We have to ask whether there is any reason why the metal–insulator transition should take place at the consolute point. This should be so only if the curve of Fig. 10.12 is symmetrical about this point. This is unlikely to be the case, and in fact the concentration at which $\sigma = \sigma_{min}$ does *decrease* with increasing temperature (Thompson 1985, p. 125).

A further assumption is that the disorder produces tails to the two Hubbard bands, which overlap, and that the transition occurs when states become delocalized at E_F, as described in Chapter 5, Section 12.

6 Dangling bonds and valence alternation pairs

Street and Mott (1975) first introduced the concept of "charged dangling bonds" (see also Mott et al. (1975) and the important developments of the theory by Kastner et al. (1976)). These are defects in chalcogenide glasses, and we apply the concept in this chapter to liquid selenium and Se–Te alloys. The concept is explained in Mott and Davis (1979) and other textbooks. In those materials, with six electrons outside a closed shell, two are inactive in s-states, two p-electrons form bonds with the two neighbours, and two p-electrons are in the so-called "lone-pair" orbitals, forming no bonds. At the end of a chain, one of the bonding p-orbitals becomes a dangling bond, with an unpaired spin. But the hypothesis made is that if this electron is removed then a lone pair on a neighbouring chain can form a strong bond with the empty orbital. There is thus threefold coordination and a positive charge; Kastner et al. denote the defect by C_3^+. At the end of an equal number of chains there will be a negative charge (C_1^-), with all orbitals either bonding or doubly charged. They have been called "valence

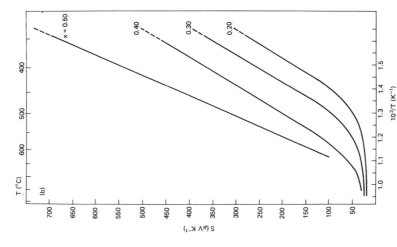

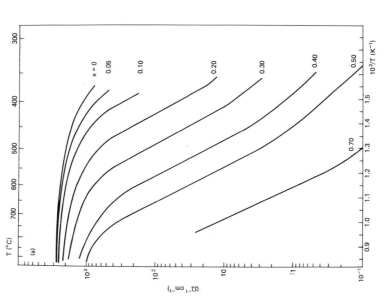

Fig. 10.22 Electrical conductivity (a) and thermoelectric power (b) of liquid Se_xTe_{1-x} alloys as functions of reciprocal temperature. From Perron (1967).

alternation pairs". These defects can lock the Fermi energy near mid-band, and have no unpaired spins.

The concept has been applied most recently by Cutler and co-workers (Fischer *et al.* 1983, Cutler 1984) to liquid selenium and selenium–tellurium alloys. These were first investigated by Perron (1967), whose data on conductivity and thermopower are shown in Figs. 10.22(a, b). It is thought that in the semiconducting liquid the Fermi energy is locked by these defects; as the temperature rises, the chain length becomes smaller, and more and more VAPs are formed, giving an approach to the threefold coordination number noted in tellurium by Cabane and Friedel (1971). In liquids Pauli paramagnetism can be caused by neutral dangling bonds, formed from VAPs by thermal excitation. Gardner and Cutler (1979) found in liquid Se–Te an activation energy 0.68 eV in the Pauli paramagnetism χ_P, and identify this as half the energy necessary to form two neutral dangling bonds from the charged pair.

The metal–insulator transition may perhaps be envisaged as similar to a band-crossing transition, caused by overlap between the C_1^- states. Since these are negatively charged, if the wave function is expressed as $\psi_a(1)\psi_b(2) + \psi_a(2)\psi_b(1)$, where ψ_a and ψ_b are both s-functions with different radii, then the larger radius could be considerable. We think that this may account for the small value of σ_{min} observed (cf. Section 4), the number of free electrons at the transition being small.

In several papers (see Kao and Cutler 1988) Cutler has pointed out that the properties of these liquids are sensitive to traces of oxygen, and has devised methods for minimizing them. This is probably because oxygen can form a bond with a chain end, thus hindering the formation of VAPs.

7   Amorphous metals; the quantum interference and interaction effects

Most amorphous metallic alloys do not show a metal–insulator transition. They do, however, show moderate changes in the resistivity with temperature, some of which can be interpreted in terms of the quantum interference effect, together with the interaction effect of Altshuler and Aronov (Chapter 5, Section 6). These will be described below. Amorphous alloys of the form Nb_xSi_{1-x}, Au_xSi_{1-x} etc. do, however, show a metal–insulator transition of Anderson type, and some of those are treated in Chapter 1, Section 7.

Amorphous calcium–aluminium alloys show a particularly large quantum interference effect. They have high resistivity, up to 400 μΩ cm ($\sigma \approx 2500 \, \Omega^{-1} \, cm^{-1}$), certainly in the Ioffe–Regel regime $l \sim a$. Howson *et al.* (1988a) found that this cannot be explained by the Ziman–Faber theory, using resistivity that do not contain a transitional metal; it is generally assumed that the high resistivity is a result of overlap of the conduction electrons into the normally empty d-states of the calcium. This seems to be confirmed by the calculations of Hafner and Jaswold (1988), who calculated the density of states of Ca_7Al_3, and found a peak in the density of states starting 1–2 eV below the Fermi energy,

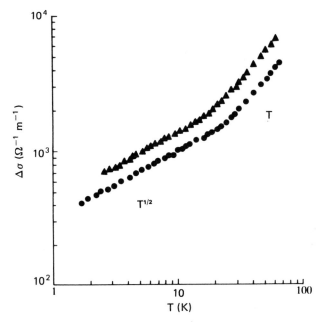

Fig. 10.23 Conductivity versus temperature for $Cu_{50}Hf_{50}$ (●) and $Cu_{50}Zr_{50}$ (▲). From Howson and Greig (1983).

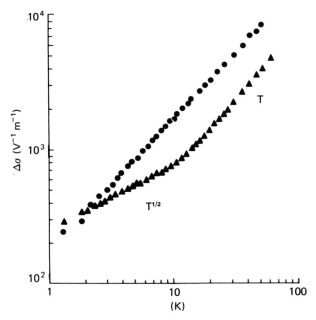

Fig. 10.24 Conductivity versus temperature for $Cu_{50}Ti_{50}$ (▲) and $Ti_{50}Be_{40}Zr_{10}$ (●). From Howson and Greig (1983).

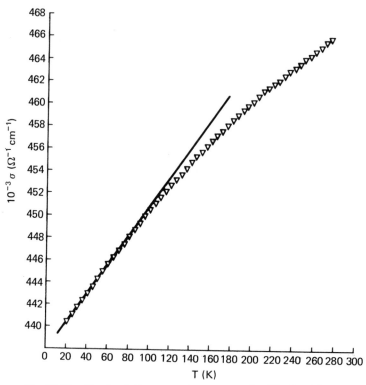

Fig. 10.25   Conductivity versus temperature for $Ti_{50}Be_{40}Zr_{10}$.

multiplying $N(E_F)$ by about 4. However, this does not seem to be confirmed by the observations of Mizutani *et al.* (1987), who found for the term γT in the low-temperature specific heat an enhancement by only 1.4. Howson *et al.* (1988a), using the equation for the quantum interference effect of Morgan *et al.* (1985), found that half the resistivity can be ascribed to it. These equations differ somewhat from ours, and indeed from others, but that the interference effect is playing a role is powerfully supported by the large negative values of $d\rho/dT$ found by Love *et al.* (1982) and Howson *et al.* (1988b). There is a linear decrease of ρ with T of about 10% between 0 and 300 K, and Howson *et al.* (1988a) found a large negative magnetoresistance. This linear term is not caused by electron–electron collisions, and is thought to be the result of collisions with phonons, in the range $T < \Theta_D$, which over a certain range gives $L_i \propto T^{-1}$ (cf. Chapter 1, Section 10).

For high-resistivity metal alloys containing transitional metals, if the conductivity is written as

$$\sigma = \sigma_0 + \Delta\sigma$$

then $\Delta\sigma$ is expected to behave like $T^{1/2}$ at low temperatures (the interaction effect: Chapter 5), then like T with increasing temperature (quantum interference with L_i caused by electron–electron interaction), and then again like $T^{1/2}$, when L_i is the result of scattering by phonons and $T \gtrsim \Theta_D$. Figures 10.23 and 10.24, for alloys for which $\sigma \approx 5 \times 10^3 \, \Omega^{-1} \text{cm}^{-1}$, show $\Delta\sigma$ in the first two regimes; Fig. 10.25 shows the reversion to $T^{1/2}$ behaviour at high T ($T > \Theta_D$).

A useful review is given by Dugdale (1987).

Belertser *et al.* (1988) have observed that the electrical resistivity of amorphous chromium films at liquid-helium temperatures jumps from a value $(10^{-3} \, \Omega \, \text{cm})$ characteristic of a poor metal by a factor 10^3, when the hydrogen content is increased sufficiently to increase the lattice constant by 10%. The transition is not abrupt, and is thought by these authors to be of Anderson type. They claim that it is the first time such a transition has been observed in a solid, and that it is similar to that in expanded mercury vapour (Section 4).

References

Abarenkov, I. V. and Heine, V. (1965) *Phil. Mag.* **12,** 529.
Abou-Chacra, R. and Thouless, D. J. (1974) *J. Phys.* **C7,** 65.
Abou-Chacra, R., Anderson, P. W. and Thouless, D. J. (1973) *J. Phys.* **C6,** 1734.
Abrahams, E., Anderson, P. W., Licciardello, D. C. and Ramakrishnan, T. W. (1979) *Phys. Rev. Lett.* **42,** 695.
Acrivos, J. V. (1972) *Phil. Mag.* **25,** 757.
Acrivos, J. V. and Mott, N. F. (1971) *Phil. Mag.* **24,** 19.
Adachi, K., Matsui, K. and Kawai, M. (1979) *J. Phys. Soc. Japan* **46,** 1474.
Adachi, K., Sato, K., Matsuura, M. and Ohashi, M. (1970) *J. Phys. Soc. Japan* **29,** 323.
Adkins, C. J. (1978) *J. Phys.* **C11,** 857.
Afonin, V. V., Galperin, Yu., M., Gurevich, V. and Schmidt, A. (1987) *Phys. Rev.* **A36,** 5729.
Afonin, V. V. and Schmidt, A. (1986) *Phys. Rev.* **B33,** 8841.
Ahmed, S., Schlenker, C. and Buder, R. (1978) *J. Magn. Magn. Mater.* **7,** 338.
Alexander, S. and Anderson, P. W. (1964) *Phys. Rev.* **133,** A1599.
Alexander, M. N. and Holcomb, D. F. (1968) *Rev. Mod. Phys.* **40,** 815.
Alexandrov, A. S., Ranninger, J. and Robaszkiewics, S. (1986) *Phys. Rev.* **B33,** 4526.
Allen, J. W. and Martin, R. M. (1980) *J. Phys. (Paris) Colloq.* **41,** C5, 17.
Altmann, S. L., Harford, A. R. and Blake, R. G. (1971) *J. Phys.* **F1,** 791.
Altshuler, B. L. and Aronov, A. G. (1979) *Solid State Commun.* **30,** 115.
Altshuler, B. L. and Aronov, A. G. (1983) *Solid State Commun.* **46,** 429.
Ambegaokar, V., Halperin, B. I. and Langer, J. S. (1971) *Phys. Rev.* **B4,** 2619.
Anderson, P. W. (1952) *Phys. Rev.* **86,** 694.
Anderson, P. W. (1956) *Phys. Rev.* **102,** 1008.
Anderson, P. W. (1958) *Phys. Rev.* **109,** 1492.
Anderson, P. W. (1961) *Phys. Rev.* **124,** 41.
Anderson, P. W. (1963) *Solid State Phys.* **14,** 99.
Anderson, P. W. (1970) *Comments Solid State Phys.* **2,** 193.
Anderson, P. W. (1973) *Mater. Res. Bull.* **8,** 153.
Anderson, P. W. (1985) *Solid State Electron.* **28,** 204.
Anderson, P. W. (1987) *Science* **235,** 1196.
Anderson, P. W., Halperin, B. I. and Varma, C. M. (1972) *Phil. Mag.* **25,** 1.
Andres, K. (1970) *Phys. Rev.* **B2,** 3768.
Andryunas, K., Vishchakas, Yu., Kabeha, V., Mochalov, I. V., Pavlyuh, A. A., Petrovskii, G. T. and Syrus, V. (1985) *Pis'ma Zh. Tekh. Fiz.* **42,** 333 [*Sov. Tech. Phys. Lett.* **11,** 108].
Appel, J. (1968) *Solid State Phys.* **21,** 193.
Asdente, M. and Friedel, J. (1961) *Phys. Rev.* **124,** 384.
Asdente, M. and Friedel, J. (1962) *Phys. Rev.* **126,** 2262.

Ashcroft, N. W. and Mermin, N. D. (1976) *Solid State Physics*. Holt, Rinehart and Winston, New York.

Ashcroft, N. W. and Russakoff, G. (1970) *Phys. Rev.* **A1**, 39.

Ashkenazi, J. and Chuchens, T. (1975) *Phil. Mag.* **32**, 763.

Ashkenazi, J. and Weger, M. (1973) *Adv. Phys.* **22**, 207.

Austin, I. G. and Mott, N. F. (1969) *Adv. Phys.* **22**, 207.

Austin, I. G. and Turner, C. E. (1969) *Phil. Mag.* **19**, 939.

Baber, W. G. (1937) *Proc. R. Soc. Lond.* **A158**, 383.

Bader, S. D., Phillips, N. E. and McWhan, D. B. (1973) *Phys. Rev.* **B7**, 4686.

Ball, M. A. (1971) *J. Phys.* **C4**, L107.

Balla, D. and Brandt, N. B. (1965) *Sov. Phys. JETP* **20**, 1111.

Banus, M. D. and Reed, T. B. (1970) *The Chemistry of Extended Defects in Non-Metallic Solids* (ed. L. Eyring and M. O'Keefe), p. 488. North-Holland, Amsterdam.

Bargeron, C. B., Avignon, M. and Drickhamer, H. G. (1971) *Inorg. Chem.* **10**, 1338.

Barlow, R. D., Jones, M., Phillips, P. J., Poynton, A. J. and Whall, T. E. (1987) *J. Appl. Phys.* **61**, 3546.

Barnes, A. C. and Enderby, J. E. (1985) *J. Non-Cryst. Solids* **77/78**, 1345.

Bass, J. (1972) *Adv. Phys.* **21**, 431.

Battles, J. W. (1971) *J. Appl. Phys.* **42**, 1280.

Baym, G. (1964) *Phys. Rev.* **135**, A1691.

Béal-Monod, M. T., Ma, S.-K. and Fredkin, D. R. (1968) *Phys. Rev. Lett.* **20**, 929.

Bednorz, J. G. and Müller, K. A. (1986) *Z. Phys.* **B64**, 189.

Beille, J., Bloch, D. and Wohlfarth, E. P. (1973) *Phys. Rev. Lett.* **43**, 207.

Belertser, B. I., Buckel, W., Hasse, J. and Brekeller, K. (1988) *Fiz. Nizk. Temp.* **14**, 578 [*Sov. J. Low Temp. Phys.* (in press)].

Berglund, G. N. and Guggenheim, H. J. (1969) *Phys. Rev.* **185**, 1022.

Bergmann, G. (1983) *Phys. Rev.* **B28**, 2914.

Bergmann, G. (1984) *Phys. Rep.* **107**, 1.

Berk, N. F. and Schrieffer, J. R. (1966) *Phys. Rev. Lett.* **17**, 433.

Besnus, R. N. and Herr, A. (1972) *Phys. Lett.* **39A**, 83.

Bhatt, R. N. and Rice, T. M. (1981) *Phys. Rev.* **B23**, 1920.

Biskupski, G. (1982) Thesis, Lille.

Biskupski, G. and Briggs, A. (1987) *J. Non-Cryst. Solids* **97/98**, 683.

Biskupski, G., Dubois, H. and Ferré, G. (1981) *Phil. Mag.* **B43**, 183.

Biskupski, G., Dubois, H., Wojkiewicz, K., Briggs, A. and Remenyi, G. (1984) *J. Phys.* **C17**, L411.

Biskupski, G., Dubois, H. and Briggs, A. (1988) *J. Phys.* **C21**, 833.

Bjerrum, N. (1951) *Kgl. Danske Vidensk. Selsk. Mat.-Fys. Medd.* **27**, 3.

Blandamer, M. J., Catterall, R., Shields, D. L. and Symonds, M. C. R. (1964) *J. Chem. Soc.* Pt IV, 4357.

Bloch, F. (1929) *Z. Phys.* **57**, 545.

Bloembergen, N. and Rowland, T. J. (1955) *Phys. Rev.* **97**, 1675.

Bogomulov, V. N., Kudinov, E. K. and Firsov, Yu. A. (1968) *Sov. Phys. Solid State* **9**, 2502.

Bosman, A. J. and Crevecoeur, C. (1966) *Phys. Rev.* **144**, 763.

Bosman, A. J. and van Daal, H. J. (1970) *Adv. Phys.* **19**, 1.

Bouchard, R. J., Perlstein, J. and Sienko, M. J. (1967) *Inorg. Chem.* **6**, 1682.

Bouchard, R. J., Gillson, J. L. and Jarrett, H. S. (1973) *Mater. Res. Bull.* **8**, 489.

Bragg, W. L. and Williams, E. J. (1934) *Proc. R. Soc. Lond.* **A145**, 699.

Brandow, B. H. (1985) *J. Solid State Chem.* **12**, 397.

Brandow, B. H. (1987) *Adv. Phys.* **20**, 651.

Bredig, M. A. (1964) *Molten Salt Chemistry* (ed. M. Blander), p. 367. Wiley, New York.

Brinkman, W. F. and Rice, T. M. (1970a) *Phys. Rev.* **B2**, 1324.

Brinkman, W. F. and Rice, T. M. (1970b) *Phys. Rev.* **B2**, 4302.

Brinkman, W. F. and Rice, T. M. (1973) *Phys. Rev.* **B7**, 1508.

Brodsky, A. M. and Tsarevsky, A. V. (1984) *J. Phys. Chem.* **88,** 3790.

Brog, K. C. and Jones, W. H. (1970) *Phys. Rev. Lett.* **24,** 58.

Bronovoi, I. L. (1980) *Zh. Eksp. Teor. Fiz.* **79,** 1936 [*Sov. Phys. JETP* **52,** 977].

Bronovoi, I. L. and Sharvin, Yu. V. (1978) *Pis'ma Zh. Eksp. Teor. Fiz.* **28,** 127 [*Sov. Phys. JETP Lett.* **28,** 117].

Brout, T. (1965) *Magnetism* (ed. G. T. Rado and H. Suhl), Vol. IIA, p. 43. Academic Press, New York.

Brusetti, R., Coey, J. M. D., Czjek, G., Fink, J., Gompf, F. and Schmidt, H. (1980) *J. Phys.* **F10,** 33.

Burow, D. F. and Lagowski, J. J. (1965) *Adv. Chem.* **50,** 125.

Buschow, K. H. J. and van Daal, H. J. (1970) *Solid State Commun.* **8,** 363.

Buyers, W. J. L., Holden, T. M., Svensen, E. C., Cowley, R. A. and Stevenson, R. W. H. (1971) *Phys. Rev. Lett.* **27,** 1442.

Cabane, B. and Friedel, J. (1971) *J. Phys. (Paris)* **32,** 73.

Callaway, J. (1964) *Energy Band Theory.* Academic Press, New York.

Callaway, J. (1987) *Phys. Rev.* **B35,** 8723.

Camphausen, B. L., Coey, J. M. D. and Chakraverty, B. K. (1976) *Phys. Rev. Lett.* **29,** 651.

Cannella, V. and Mydosh, J. R. (1972) *Phys. Rev.* **B6,** 4220.

Cannella, V., Mydosh, J. R. and Budnick, J. I. (1971) *J. Appl. Phys.* **42,** 1689.

Caplin, A. D. and Rizzuto, C. (1968) *Phys. Rev. Lett.* **21,** 746.

Care, C. M. and March, N. H. (1971) *J. Phys.* **C4,** L372.

Caroli, B. (1967) *J. Phys. Chem. Solids* **28,** 1427.

Castner, T. G. (1980) *Phil. Mag.* **B42,** 643.

Catterall, R. (1970) *Phil. Mag.* **22,** 779.

Catterall, R. and Mott, N. F. (1969) *Adv. Phys.* **18,** 605.

Cauchois, Y. and Mott, N. F. (1949) *Phil. Mag.* **40,** 1260.

Chandrasekhar, G. V., Won Choi, Q., Moyo, J. and Honig, J. M. (1970) *Mater. Res. Bull.* **5,** 949.

Chao, K. A. (1971) *Phys. Rev.* **B4,** 4034.

Chao, K. A. and Gutzwiller, M. C. (1971) *J. Appl. Phys.* **42,** 1420.

Chen, H. L. S. and Sladek, R. J. (1978) *Phys. Rev.* **B18,** 6824.

Chieux, P. and Sienko, M. J. (1970) *J. Chem. Phys.* **53,** 566.

Chouteau, G., Fourneaux, R., Gobrecht, K. and Tournier, R. (1968a) *Phys. Rev. Lett.* **20,** 193.

Chouteau, G., Fourneaux, R., Tournier, R. and Lederer, P. (1968b) *Phys. Rev. Lett.* **21,** 1082.

Clogston, H. M., Matthias, B. T., Peter, M. I., Williams, H. J., Corenzwit, E. and Sherwood, R. C. (1962) *Phys. Rev.* **125,** 541.

Coey, J. M. D., Brusetti, R., Kallel, A., Schweizer, J. and Fuess, H. (1974) *Phys. Rev. Lett.* **32,** 1257.

Cohen, M. H. and Jortner, J. (1973) *Phys. Rev. Lett.* **30,** 699.

Cohen, M. H. and Thompson, J. C. (1968) *Adv. Phys.* **17,** 857.

Cohen, M. H., Fritzsche, H. and Ovshinsky, S. R. (1969) *Phys. Rev. Lett.* **22,** 1063.

Cohen, M. L. (1982) *Physica Scripta (Sweden)* **T1,** 5.

Coles, B. R. (1952) *Proc. Phys. Soc.* **B65,** 221.

Coles, B. R. (1987) *Contemp. Phys.* **28,** 143.

Coles, B. R. and Taylor, J. C. (1962) *Proc. R. Soc. Lond.* **A207,** 139.

Collings, E. W. (1971) *J. Phys. (Paris) Colloq.* **32,** C1, 516.

Conwell, E. M. (1956) *Phys. Rev.* **103,** 57.

Coqblin, B. and Blandin, A. (1968) *Adv. Phys.* **7,** 281.

Cottrell, A. H. (1988) *Introduction to the Modern Theory of Metals.* Institute of Metals, London.

Crevecoeur, C. and de Witt, H. J. (1968) *Solid State Commun.* **6,** 295.

Cullen, J. R. and Callen, E. (1971) *Phys. Rev. Lett.* **26,** 236.

Cullen, J. R. and Callen, E. (1973) *Phys. Rev.* **B7**, 397.

Cullen, J. R., Callen, E. and Luther, A. H. (1968) *Phys. Rev.* **170**, 733.

Cutler, M. (1977) *Liquid Semiconductors.* Academic Press, New York.

Cutler, M. (1984) *Phil. Mag.* **B49**, 83.

Cutler, M. and Field, M. A. (1968) *Phys. Rev.* **169**, 632.

Cutler, M. and Leavy, J. F. (1964) *Phys. Rev.* **133**, A1153.

Cutler, M. and Mott, N. F. (1969) *Phys. Rev.* **181**, 1336.

Cyrot, M. (1987) *Solid State Commun.* **63**, 1013.

Damay, P. (1973) Thesis, University of Paris.

Damay, P., Depoorter, M., Chieux, P. and Lepoutre, G. (1969) *Metal–Ammonia Solutions* (ed. J. J. Lagowski and M. J. Sienko), p. 233. Butterworth, London.

Damay, P., Leclercq, F. and Lelieur, J. P. (1988) *Phil. Mag.* **B57**, 75.

Davies, J. H. (1980) *Phil. Mag.* **B41**, 370.

Davies, J. H., Lee, P. and Rice, T. M. (1982) *Phys. Rev. Lett.* **49**, 7500.

Davis, E. A. and Compton, W. D. (1965) *Phys. Rev.* **A140**, 2183.

de Boer, F. R., Schinkel, C. J., Biesterbos, J. and Proost, S. (1969) *J. Appl. Phys.* **40**, 1049.

de Boer, J. H. and Verwey, E. J. W. (1937) *Proc. Phys. Soc.* **A49**, 59.

de Chatel, P. F. and Wohlfarth, E. A. (1973) *Comments Solid State Phys.* **5**, 133.

de Dood, W. and de Chatel, P. F. (1973) *J. Phys.* **F3**, 1039.

de Gennes, P. G. (1960) *Phys. Rev.* **118**, 141.

de Gennes, P. G. (1962) *J. Phys. Radium* **23**, 630.

de Gennes, P. G. and Friedel, J. (1958) *J. Phys. Chem. Solids* **4**, 71.

de Haas, W. J. and de Boer, J. H. (1934) *Physica* **1**, 605.

Dernier, P. D. and Marezio, M. (1970) *Phys. Rev.* **B2**, 3771.

Dersch, U. and Thomas, P. (1985) *J. Phys.* **C18**, 5815.

Dersch, U., Overhof, H. and Thomas, P. (1987) *J. Non-Cryst. Solids* **97/98**, 9.

Devreese, J. T. (1972) *Polarons in Ionic Crystals and Polar Semiconductors.* North-Holland, Amsterdam.

di Castro, C. (1988) *Anderson Localization* (ed. T. Ando and H. Fukuyama), p. 119. Springer-Verlag, Berlin.

Donally, J. M. and Cutler, M. (1972) *J. Phys. Chem. Solids* **33**, 1017.

Doniach, S. and Engelsberg, S. (1966) *Phys. Rev. Lett.* **17**, 750.

Dougier, P. and Casalot, A. (1970) *J. Solid State Chem.* **2**, 396.

Dougier, P. and Hagenmuller, P. (1975) *J. Solid State Chem.* **15**, 158.

Doumerc, J. P. (1978) *The Metal–Non-Metal Transition in Disordered Systems: Proceedings of 19th Scottish Universities Summer School in Physics* (ed. L. R. Friedman and D. P. Tunstall), p. 313. SUSSP, Edinburgh.

Doumerc, J. P., Pouchard, M. and Hagenmuller, P. (1985) *The Metallic and Non-Metallic States of Matter* (ed. P. P. Edwards and C. N. R. Rao), p. 287. Taylor & Francis, London.

Dreirach, O., Evans, R., Gunterodt, H. J. and Kunzi, H. U. (1972) *J. Phys.* **F2**, 709.

Drickamer, H. G. and Frank, C. W. (1973) *Electronic Transitions and High Pressure Chemistry and Physics of Solids.* Chapman & Hall, London.

Drickamer, H. G., Lynch, R. W., Clenenden, R. L. and Perez-Albuerne, E. A. (1966) *Solid State Phys.* **19**, 135.

Dubois, H., Biskupski, G., Spriet, J. B. and Briggs, H. (1985) *J. Phys.* **C18**, L195.

Dubson, M. A. and Holcomb, D. F. (1985) *Phys. Rev.* **B32**, 1955.

Dugdale, J. S. (1987) *Contemp. Phys.* **28**, 547.

Dugdale, J. S. and Guénault, A. M. (1966) *Phil. Mag.* **13**, 563.

Dumas, J. (1978) Thesis, Grenoble.

Dumas, J. and Schlenker, C. (1976) *J. Phys. (Paris) Colloq.* **37**, C4, 41.

Dumas, J. and Schlenker, C. (1979a) *Phys. Rev.* **B20**, 3913.

Dumas, J. and Schlenker, C. (1979b) *J. Phys.* **C12**, 2361.

Durkan, J., Elliott, R. J. and March, N. H. (1968) *Rev. Mod. Phys.* **40**, 812.

Dunlop, J. B., Grüner, G. and Napoli, F. (1984) *Solid State Commun.* **15**, 13.

Dye, J. L., Sankauer, R. F. and Smith, G. E. (1960) *J. Am. Chem. Soc.* **83**, 4797.

Eagles, D. M. (1969a) *Phys. Rev.* **178**, 668.

Eagles, D. M. (1969b) *Phys. Rev.* **181**, 1278.

Eagles, D. M. (1984) *J. Phys.* **C17**, 637.

Eagles, D. M. (1985) *Physics of Disordered Materials: Mott Festschrift*, Vol. 1 (ed. D. Adler, H. Fritzsche and S. R. Ovshinsky), p. 375. Plenum Press, New York.

Eagles, D. M. and Lalouse, P. (1984) *J. Phys.* **C17**, 699.

Economou, E. N. (1972) *Phys. Rev. Lett.* **28**, 1206.

Economou, E. N. and Cohen, M. H. (1972) *Phys. Rev.* **B5**, 2931.

Economou, E. N., Soukoulis, L. M., Cohen, M. H. and Zdetzis, A. D. (1985) *Phys. Rev.* **B31**, 6172.

Edwards, D. M. (1979) *Electrons in Disordered Materials and at Metallic Surfaces* (ed. P. Phariseau and B. L. Gyorffy), p. 353. Plenum Press, New York.

Edwards, D. M. and Wohlfarth, E. P. (1968) *Proc. R. Soc. Lond.* **A303**, 127.

Edwards, J. T. and Thouless, D. J. (1972) *J. Phys.* **C5**, 807.

Edwards, P. P. and Sienko, M. J. (1978) *Phys. Rev.* **B17**, 2575.

Edwards, P. P. and Sienko, M. J. (1983) *Int. Rev. Phys. Chem.* **3**, 83.

Edwards, S. F. (1958) *Phil. Mag.* **3**, 1020.

Edwards, S. F. (1961) *Phil. Mag.* **6**, 617.

Edwards, S. F. (1962) *Proc. R. Soc. Lond.* **A267**, 578.

Edwards, S. F. and Hillel, A. J. (1968) *J. Phys.* **C1**, 61.

Efros, A. L. and Shklovskii, B. L. (1975) *J. Phys.* **C8**, L49.

Elliott, R. J. (1965) *Magnetism* (ed. G. T. Rado and H. Suhl), Vol. IIA, p. 385. Academic Press, New York.

Elliott, R. J. and Heap, B. R. (1962) *Proc. R. Soc. Lond.* **A265**, 264.

Elyutin, P. V., Hickey, B., Morgan, G. J. and Weir, G. F. (1984) *Phys. Stat. Solidi* **(b)124**, 279.

Emery, V. J. (1987) *Phys. Rev. Lett.* **58**, 2794.

Emin, D. (1973) *Adv. Phys.* **22**, 57.

Emin, D. (1975) *Adv. Phys.* **24**, 304.

Emin, D. (1977a) *Phil. Mag.* **35**, 89.

Emin, D. (1977b) *Solid State Commun.* **22**, 409.

Emin, D. and Hillery, M. S. (1989) *Phys. Rev.* **B39**, 6575.

Enderby, J. E. and Barnes, A. C. (1990) *Rep. Prog. Phys.* **53**, 85.

Endo, S., Mitsui, T. and Mihadai, T. (1973) *Phys. Lett.* **46A**, 29.

Eskes, H. and Sawatzky, G. A. (1989) *Phys. Rev. Lett.* **61**, 1415.

Evans, R. (1970) *J. Phys.* **C3**, S137.

Evans, R., Greenwood, D. A. and Lloyd, P. (1971) *Phys. Lett.* **35**, 57.

Faber, T. E. (1972) *Introduction to the Theory of Liquid Metals.* Cambridge University Press.

Falicov, L. M. and Kimball, J. C. (1969) *Phys. Rev. Lett.* **22**, 997.

Fan, J. C. C. (1972) Thesis, Harvard University.

Fazekas, P. (1972) *Solid State Commun.* **10**, 175.

Ferré, D., Dubois, H. and Biskupski, G. (1975) *Phys. Stat. Solidi* **(b)70**, 81.

Ferrier, R. P. and Herrell, D. J. (1969) *Phil. Mag.* **19**, 853.

Ferrier, R. P. and Herrell, D. J. (1970) *J. Non-Cryst. Solids* **2**, 278.

Fibich, M. and Ron. A. (1970) *Phys. Rev. Lett.* **25**, 296.

Finkelstein, A. M. (1983) *Sov. Phys. JETP* **57**, 97.

Fischer, R., Cutler, M. and Rasolondramanitra, H. (1983) *Phil. Mag.* **B48**, 537.

Fluitman, J. H. J., Boom, R., de Chatel, P. F., Schenkel, C. J., Tilanus, J. L. L. and de Vries, B. R. (1973) *J. Phys.* **F3**, 109.

Föex, M. (1946) *C.R. Acad. Sci. Paris* **B223**, 1126.

Fogle, W. and Perlstein, J. H. (1972) *Phys. Rev.* **B6**, 1402.

Franz, J. and Davies, J. H. (1986) *Phys. Rev. Lett.* **57**, 475.

Frederikse, H. P. R., Thurber, W. R. and Hosler, R. (1964) *Phys. Rev.* **A134**, 442.

Freed, K. F. (1972) *Phys. Rev.* **B5**, 4802.

Freed, K. F. and Sugarman, N. (1943) *J. Chem. Phys.* **11**, 354.

Frenkel, J. (1931) *Phys. Rev.* **37**, 17.

Freyland, W. and Hensel, F. (1985) *The Metallic and Non-Metallic States of Matter* (ed. P. P. Edwards and C. N. R. Rao), p. 93. Taylor & Francis, London.

Freyland, W., Pfeifer, H. P. and Hensel, F. (1974) *Proceedings of 5th Conference on Amorphous and Liquid Semiconductors* (ed. J. Stuke and W. Brenig), p. 1327. Taylor & Francis, London.

Friedel, J. (1952a) *Phil. Mag.* **43**, 153.

Friedel, J. (1952b) *Phil. Mag.* **43**, 1115.

Friedel, J. (1954) *Adv. Phys.* **3**, 446.

Friedel, J. (1956) *Can. J. Phys.* **34**, 1190.

Friedel, J. (1969) *Comments Solid State Phys.* **2**, 29.

Friedel, J. (1973) *J. Phys.* **F3**, 785.

Friedel, J., Léman, G. and Olszewski, S. (1961) *J. Appl. Phys. Suppl.* **32**, 3258.

Friedel, J., Lenglart, P. and Léman, G. (1964) *J. Phys. Chem. Solids* **25**, 781.

Friedman, L. (1971) *J. Non-Cryst. Solids* **62**, 329.

Friedman, L. and Holstein, T. (1963) *Ann. Phys.* **21**, 474.

Friend, R. H. and Jérome, D. (1979) *J. Phys.* **C12**, 1441.

Friend, R. H. and Yoffe, A. D. (1987) *Adv. Phys.* **36**, 1.

Fritsch, G., Schulte, A. and Lüscher, E. (1987) *Amorphous and Liquid Materials* (ed. E. Lüscher, G. Fritsch and G. Macucci), p. 368. Plenum Press, New York.

Fritzsche, H. (1958) *J. Phys. Chem. Solids* **6**, 69.

Fritzsche, H. (1959) *Phys. Rev.* **115**, 336.

Fritzsche, H. (1960) *Phys. Rev.* **119**, 1899.

Fritzsche, H. (1978) *The Metal–Non-Metal Transition in Disordered Systems: Proceedings of the 19th Scottish Universities Summer School in Physics* (ed. L. R. Friedman and D. P. Tunstall), p. 193. SUSSP, Edinburgh.

Fritzsche, H. and Cuevas, M. (1960) *Phys. Rev.* **119**, 1238.

Fröhlich, H. (1950) *Phys. Rev.* **79**, 845.

Fuchs, R. (1965) *J. Chem. Phys.* **42**, 3781.

Fukuyama, H. and Hoshino, K. (1981) *J. Phys. Soc. Japan* **50**, 2131.

Fukuyama, H. and Nagai, T. (1971a) *Phys. Rev.* **B3**, 4413.

Fukuyama, H. and Nagai, T. (1971b) *J. Phys. Soc. Japan* **31**, 812.

Fulda, P. (1988) *J. Phys.* **F18**, 601.

Fulda, P., Keller, J. and Kwichnagel, G. (1988) *Solid State Phys.* **41**, 1.

Furdyna, J. K. and Kossut, J. (1988) *Semiconductors and Semi-Metals*, Vol. 25: *Diluted Magnetic Semiconductors*, Chap. 10. Academic Press, New York.

Gallager, W. J. (1988) *J. Appl. Phys.* **63**, 4216.

Gardner, J. A. and Cutler, M. (1977) *Proceedings of 7th International Conference on Amorphous and Liquid Semiconductors* (ed. W. E. Spear), p. 838. Centre for Industrial Consultancy, University of Edinburgh.

Gaspard, J. P. and Cyrot-Lackmann, F. (1972) *J. Phys.* **C5**, 3047.

Gaspard, J. P. and Cyrot-Lackmann, F. (1973) *J. Phys.* **C6**, 3077.

Gill, J. C. (1986) *Contemp. Phys.* **27**, 37.

Glaunsinger, W. S. Zolotov, S. and Sienko, M. J. (1972) *J. Chem. Phys.* **56**, 4756.

Goncalvez da Silva, C. E. T. and Falicov, L. M. (1972) *J. Phys.* **C5**, 908.

Goodenough, J. B. (1963) *Magnetism and the Chemical Bond*. Wiley, New York.

Goodenough, J. B. (1971) *Prog. Solid State Chem.* **5**, 145.

Goodenough, J. B. (1972) *Phys. Rev.* **B5**, 2764.

Goodenough, J. B., Mott, N. F., Pouchard, M., Demazeau, G. and Hagenmuller, P. (1973) *Mater. Res. Bull.* **8**, 647.

Gossard, A. C., McWhan, D. B. and Remeika, J. P. (1970) *Phys. Rev.* **B2**, 3762.

Gosselin, J. R., Townsend, M. G., Tremblay, R. J., Repley, L. G. and Carson, D. W. (1973) *J. Phys.* **C6**, 1661.

Götze, W. (1979) *J. Phys.* **C12**, 1279.

Götze, W. (1981) *Phil. Mag.* **B43**, 219.

Gourmala, M., Fouracdot, G. and Mercier, J. (1979) *J. Crystal Growth* **46**, 132.

Graener, H., Rosenberg, M., Whall, T. E. and Jones, M. R. B. (1979) *Phil. Mag.* **B40**, 389.

Green, M. P., Aldridge, C. and Bajaj, K. K. (1977) *Phys. Rev.* **B15**, 2217.

Greene, M. and Kohn, W. (1965) *Phys. Rev.* **137**, 513.

Greenwood, D. A. (1958) *Proc. Phys. Soc. Lond.* **71**, 585.

Gruber, O. F. and Gardner, J. A. (1971) *Phys. Rev.* **B4**, 3994.

Grüner, G. and Mott, N. F. (1974) *J. Phys.* **F4**, L10.

Grüner, J. D., Shanks, H. R. and Wallace, D. C. (1962) *J. Chem. Phys.* **30**, 772.

Gutzwiller, M. C. (1963) *Phys. Rev. Lett.* **10**, 159.

Gutzwiller, M. C. (1964) *Phys. Rev.* **134**, A923.

Hafner, J. and Jaswold, S. S. (1988) *J. Phys.* **F18**, L3.

Hagenmuller, P. (1971) *Prog. Solid State Chem.* **5**, 71.

Halperin, B. I. and Rice, T. M. (1968a) *Rev. Mod. Phys.* **40**, 755.

Halperin, B. I. and Rice, T. M. (1968b) *Solid State Phys.* **21**, 115.

Hamann, D. R., Schluter, M. and Chiang, C. (1979) *Phys. Rev. Lett.* **43**, 1494.

Harrison, W. A. (1966) *Pseudopotentials in the Theory of Metals.* Benjamin, New York.

Hasegawa, R. and Tsuei, C. C. (1970) *Phys. Rev.* **B2**, 1631.

Hedin, L. and Lundqvist, S. (1969) *Solid State Phys.* **23**, 1.

Heeger, A. J. (1969) *Solid State Phys.* **23**, 283.

Heikes, R. R. (1961) *Thermoelectricity* (ed. R. R. Heikes and R. W. Ure), p. 80. Interscience, New York.

Heine, V. (1969) *The Physics of Metals* (ed. J. M. Ziman), p. 1. Cambridge University Press.

Heine, V. (1970) *Solid State Phys.* **24**, 1.

Hensel, F. and Franck, E. U. (1968) *Rev. Mod. Phys.* **40**, 695.

Hensel, J. C., Phillips, T. G. and Thomas, G. A. (1977) *Solid State Phys.* **32**, 87.

Herrell, D. J. (1970) PhD Thesis, University of Cambridge.

Herring, C. (1966) *Magnetism* (ed. G. T. Rado and H. Suhl), Vol. IV. Academic Press, New York.

Hertel, G., Bishop, D. J., Spencer, E. G., Royel, J. M. and Dynes, R. C. (1983) *Phys. Rev. Lett.* **50**, 974.

Herzfeld, K. F. (1927) *Phys. Rev.* **29**, 701.

Hickami, S. (1981) *Phys. Rev.* **B24**, 2687.

Hindley, N. K. (1970) *J. Non-Cryst. Solids* **5**, 17 and 31.

Hirsch, M. J. and Holcomb, D. F. (1987) *Disordered Semiconductors: H. Fritzsche Festschrift* (ed. M. A. Kastner, G. A. Thomas and S. R. Ovshinsky), p. 45. Plenum Press, New York.

Hodges, C., Smith, H. and Wilkins, J. W. (1971) *Phys. Rev.* **B4**, 302.

Hollinger, G., Pertosa, P., Doumerc, J., Himpsel, F. J. and Reihl, B. (1985) *Phys. Rev.* **B32**, 1987.

Holstein, T. (1959) *Ann. Phys. (NY)* **8**, 343.

Holstein, T. and Friedman, L. R. (1968) *Phys. Rev.* **165**, 1019.

Honig, J. M. (1984) *Basic Properties of Binary Oxides* (ed. A. Dominquez Rodrigues, J. Caistay and R. Marques), p. 101. Seville University Press.

Honig, J. M. (1985) *The Metallic and Non-Metallic States of Matter* (ed. P. P. Edwards and C. R. N. Rao), p. 261. Taylor & Francis, London.

Hopfield, J. J. (1969) *Comments Solid State Phys.* **2**, 40.

Hopkins, P. F., Burns, M. J. Rimberg, A. J. and Westerveld, R. M. (1989) *Phys. Rev.* **B38**, 1660.

Howat, B. and Webster, B. C. (1972) *J. Phys. Chem.* **70**, 3714.

Howson, M. A. and Greig, D. (1983) *J. Phys.* **F13**, L155.

Howson, M. A., Hickey, B. J. and Morgan, G. J. (1988a) *Phys. Rev.* **B36**, 5267.

Howson, M. A., Paja, A., Morgan, G. J. and Walker, M. J. (1988b) *Z. Phys. Chem.* **157**, 693.

Hubbard, J. (1964a) *Proc. R. Soc. Lond.* **A227**, 237.

Hubbard, J. (1964b) *Proc. R. Soc. Lond.* **281**, 401.

Huefner, S. (1985) *Solid State Commun.* **53**, 707.

Hung, C. S. and Gleissmann, J. R. (1950) *Phys. Rev.* **79**, 726.

Huster, E. (1938) *Annln Phys.* **33**, 477.

Ihle, D. and Lorenz, B. (1980) *Phil. Mag.* **B42**, 337.

Ihle, D. and Lorenz, B. (1985) *J. Phys.* **C18**, L647.

Imry, Y. and Gefen, Y. (1984) *Phil. Mag.* **B50**, 203.

Imry, Y., Gefen, Y. and Bergman, D. J. (1982) *Phys. Rev.* **B26**, 3430.

Ioffe, A. F. and Regel, A. R. (1960) *Prog. Semicond.* **4**, 237.

Ionov, A. N., Matveev, M. N., Rentch, R. and Shlimak, I. S. (1985) *Pis'ma Zh. Eksp. Teor. Fiz.* **42**, 330 [*Sov. Phys. JETP Lett.* **42**, 466 (1986)].

Itskevich, E. S. and Fisher, L. M. (1964) *Sov. Phys. JETP* **26**, 66.

Jaccarino, V. and Walker, L. R. (1965) *Phys. Rev. Lett.* **15**, 258.

Jarrett, H. S., Cloud, W. H., Bouchard, R. J., Butler, S. R., Frederick, C. G. and Gillson, J. L. (1968) *Phys. Rev. Lett.* **21**, 617.

Jarrett, H. S., Bouchard, R. J., Gillson, J. L., Jones, G. A., Marcus, S. M. and Weiter, J. F. (1973) *Mater. Res. Bull.* **8**, 877.

Jayaraman, A. (1972) *Phys. Rev. Lett.* **29**, 1674.

Jayaraman, A., McWhan, D. B., Remeika, J. P. and Dernier, P. D. (1970a) *Phys. Rev.* **B2**, 3751.

Jayaraman, A., Narayanamurti, V., Bucher, F. and Maines, R. G. (1970b) *Phys. Rev. Lett.* **25**, 368 and 1480.

Jérome, D., Ryter, D., Shultz, H. J. and Friedel, J. (1985) *Phil. Mag.* **B52**, 403.

Johansen, G. and Mackintosh, A. R. (1970) *Solid State Commun.* **8**, 121.

Jones, H. (1960) *The Theory of Brillouin Zones and Electronic States in Crystals.* North-Holland, Amsterdam [2nd edn 1971].

Jones, H. and Mott, N. F. (1937) *Proc. R. Soc. Lond.* **A162**, 49.

Jones, H., Mott. N. F. and Skinner, H. W. B. (1934) *Phys. Rev.* **45**, 379.

Jortner, J. (1959) *J. Chem. Phys.* **30**, 839.

Joshi, G. M., Pai, M., Harrison, H., Sandberg, G. J., Aragon, R. and Honig, J. M. (1980) *Mater. Res. Bull.* **15**, 1515.

Jullien, R. and Jérome, D. (1971) *J. Phys. Chem. Solids* **32**, 257.

Kahn, D. C. and Erickson, R. A. (1970) *Phys. Rev.* **B1**, 2243.

Kamimura, H., Matsumo, S. and Saito, R. (1988) *Solid State Commun.* **67**, 363.

Kanamori, J. (1963) *Prog. Theor. Phys.* **30**, 275.

Kao, S. S. and Cutler, M. (1988) *Phys. Rev.* **B37**, 581.

Kaplan, T. A., Mahaute, S. D. and Hartman, W. M. (1971) *Phys. Rev. Lett.* **27**, 1796.

Kasowski, R. V. (1973) *Phys. Rev.* **B8**, 1378.

Kastner, M., Adler, D. and Fritzsche, H. (1976) *Phys. Rev. Lett.* **37**, 1504.

Kasuya, T. (1956) *Prog. Theor. Phys.* **16**, 45.

Kasuya, T. (1970) *Solid State Commun.* **8**, 1035.

Kasuya, T. and Koida, S. (1958) *J. Phys. Soc. Japan* **3**, 1287.

Katsumoto, S., Komori, F., Sano, N. and Kobayashi, S. (1988) *J. Phys. Soc. Japan* **56**, 2257.

Kaveh, M. (1984) *Phil. Mag.* **B50**, 175.

Kaveh, M. (1985a) *Phil. Mag.* **B52**, L1.

Kaveh, M. (1985b) *Phil. Mag.* **B52**, 41.

Kaveh, M. (1985c) *Phil. Mag.* **B52**, 521.

Kaveh, M. (1985d) *J. Phys.* **C17**, L79.

Kaveh, M. and Liebert, A. (1988) *Phil. Mag. Lett.* **58**, 247.

Kaveh, M. and Mott, N. F. (1981a) *J. Phys.* **C14**, L177.
Kaveh, M. and Mott, N. F. (1981b) *J. Phys.* **C14**, L183.
Kaveh, M. and Mott, N. F. (1982) *J. Phys.* **C15**, L697 and L707.
Kaveh, M. and Mott, N. F. (1987) *Phil. Mag.* **B55**, 1 and 9.
Kaveh, M., Newson, D. J., Ben Zimra, D. and Pepper, M. (1987) *J. Phys.* **C20**, L19.
Kaveh, M. and Wiser, N. (1984) *Adv. Phys.* **33**, 257.
Kawabata, A. (1981) *Solid State Commun.* **38**, 823.
Kawabata, A. (1984) *J. Phys. Soc. Japan* **53**, 1429.
Kawabata, A. (1988) *Anderson Localization* (ed. T. Ando and H. Fukuyama), p. 116.
    Springer-Verlag, Berlin.
Kawai, N. and Mochizuki, S. (1971) *Solid State Commun.* **9**, 1393.
Kawamoto, H., Honig, J. M. and Appel, J. (1980) *Phys. Rev.* **B22**, 2626.
Kawano, S., Kosuge, G. and Kachi, S. (1966) *J. Phys. Soc. Japan* **21**, 2744.
Keldysh, L. V. (1986) *Contemp. Phys.* **27**, 395.
Keldysh, L. V. and Kopaev, Yu. V. (1965) *Sov. Phys. Solid State* **6**, 2219.
Kerlin, A. I., Nagasawa, H. and Jérome, D. (1973) *Solid State Commun.* **13**, 1125.
Kevan, L. and Webster, B. (1976) *Electron Solvent and Anion Solvent Interactions.* Elsevier,
    Amsterdam.
Kirk, J. L., Vedam, K., Narayanamurti, V., Jayaraman, A. and Bucher, E. (1972) *Phys. Rev.*
    B6, 3023.
Kirkpatrick, S. (1973) *Rev. Mod. Phys.* **45**, 574.
Kiwi, M. and Ramirez, R. (1972) *Phys. Rev.* **B6**, 3700.
Klein, A. P. and Heeger, A. J. (1966) *Phys. Rev.* **144**, 458.
Klein, M. W. and Brout, T. (1963) *Phys. Rev.* **132**, 2412.
Klipstein, P. C., Friend, R. H. and Yoffe, A. D. (1985) *Phil. Mag.* **B52**, 611.
Knapp, G. S., Fradin, F. Y. and Culbert, H. V. (1971) *J. Appl. Phys.* **42**, 1341.
Knox, R. S. (1963) *Theory of Excitons* (*Solid State Phys.* Supp. No. 5), p. 207. Academic
    Press, New York.
Kohn, W. (1967) *Phys. Rev. Lett.* **19**, 439.
Kolber, M. A. and MacCrone, A. K. (1972) *Phys. Rev. Lett.* **29**, 1457.
Kouvel, J. S. and Comly, J. D. (1970) *Phys. Rev. Lett.* **24**, 598.
Kramer, B., MacKinnon, A. and Weaire, D. (1981) *Phys. Rev.* **23**, 6357.
Kramhansl, J. A. (1965) *Physics of Solids at High Pressures* (ed. C. T. Tomizuka and R. M.
    Emrich), p. 425. Academic Press, New York.
Kraus, C. A. (1921) *J. Am. Chem. Soc.* **43**, 749.
Kreiger, J. B. and Nightingale, M. (1971) *Phys. Rev.* **B4**, 1266.
Krill, G., Pannissod, P., La Pierre, M. F., Gautier, F., Robert, C. and Eddine, M. (1976)
    J. Phys. **C9**, 152.
Kubo, R. (1956) *Can. J. Phys.* **34**, 1274.
Kugler, A. A. (1969) *Ann. Phys. (NY)* **53**, 133.
Kuiper, P., Kruizinga, G., Ghijzen, J., Sawatzky, G. A. and Verweij, H. (1989) *Phys. Rev.*
    Lett. **62**, 221.
Kuwamoto, H., Honig, J. M. and Appel, J. (1980) *Phys. Rev.* **B22**, 2626.
Kwizera, P., Mabatah, A. K., Dresselhaus, M. S. and Adler, D. (1981) *Phys. Rev.* **B24**, 2972.
Lakkis, S., Schlenker, C., Chakraverty, B. K., Buder, R. and Marezio, M. (1976) *Phys. Rev.*
    B14, 1429.
Landau, L. D. (1933) *Phys. Z. Sowjetunion* **3**, 669.
Landau, L. D. (1957) *Sov. Phys. JETP* **3**, 920.
Landau, L. D. and Pomeranchuk, I. (1937) *Phys. Z. Sowjetunion* **10**, L49.
Landau, L. D. and Zeldovich, G. (1943) *Acta Phys. Chem. USSR* **18**, 194.
Langer, J. S. and Ambegaokar, V. (1961) *Phys. Rev.* **121**, 1090.
Langer, W., Plishke, M. and Mattis, D. (1969) *Phys. Rev. Lett.* **23**, 1448.
Laredo, E., Rowan, L. J. and Slifkin, Z. (1981) *Phys. Rev. Lett.* **47**, 384.
Laredo, E., Paul, W. B., Rowan, L. and Slifkin, L. (1983) *Phys. Rev.* **B27**, 2470.

Leclercq, F. and Damay, P. (1988) *Phil. Mag.* **B57,** 6.

Lederer, P. and Mills, D. L. (1968) *Phys. Rev.* **165,** 837.

Lederer, P., Launois, H., Pouget, J. P., Casalot, A. and Villeneuve, G. (1972) *J. Phys. Chen Solids* **33,** 1969.

Lee, P. A. (1982) *Anderson Localization* (ed. Y. Nagaoka and H. Fukuyama), p. 7ε Springer-Verlag, Berlin.

Lemos, V., Gualberto, G. M., Salzberg, J. B. and Cerdiera, F. (1980) *Phys. Stat. Solia* **(b)100,** 755.

Liang, V. K. C. and Tsuei, L. C. (1973) *Phys. Rev.* **B7,** 3215.

Liang, W. Y. (1989) Private communication.

Licciardello, D. C., Di Marco, R. and Economou, E. N. (1977) *Phys. Rev.* **B24,** 543.

Lifshitz, I. M. (1964) *Adv. Phys.* **13,** 483.

Lightsey, P. A. (1973) *Phys. Rev.* **B8,** 3586.

Lightsey, P. A. and Holcomb, D. F. (1976) *Phys. Rev.* **B13,** 1821.

Logan, D. E. and Edwards, P. P. (1985) *The Metallic and Non-Metallic States of Matter* (ed P.P. Edwards and C. R. N. Rao), p. 65. Taylor & Francis, London.

Long, A. P. and Pepper, M. (1984) *J. Phys.* **C17,** 3391.

Lonzarich, G. G. and Taillefer, L. (1985) *J. Phys.* **C18,** 4339.

Loram, J. W., Whall, T. E. and Ford, P. J. (1971) *Phys. Rev.* **B3,** 953.

Lorenz, B. and Ihle, D. (1972) *Phys. Stat. Solidi* **(b)54,** 463.

Lourde, R. D. and Windsor, G. C. (1970) *Adv. Phys.* **19,** 813.

Love, D. P., Wang, P. C., Naugle, D. G., Tsai, C. L., Geissen, B. C. and Calloway, P. C (1982) *Phys. Lett.* **90A,** 303.

Lucovsky, G. (1978) *Bull. Am. Phys. Soc.* **23,** 373.

Lucovsky, G., Allen, J. W. and Allen, R. (1979) *Physics of Semiconductors*, p. 465. Institute of Physics, London.

Ludwig, H. (1985) *Localization and Metal–Insulator Transitions* (ed. H. Fritzsche anc D. Adler), p. 90. Plenum Press, New York.

Ludwig, R. and Micklitz, H. (1984) *Solid State Commun.* **50,** 861.

Ludwig, R., Razavi, F. S. and Micklitz, H. (1981) *Solid State Commun.* **39,** 361.

Lukes, T. (1972) *J. Non-Cryst. Solids* **8/10,** 470.

Luttinger, J. M. (1960) *Phys. Rev.* **119,** 1153.

Lyo, S. K. (1972) *Phys. Rev. Lett.* **28,** 1192.

Mabatah, A. K., Yoffa, E. J., Ecklund, P. C., Dresselhaus, M. S. and Adler, D. (1980) *Phys Rev.* **B21,** 1676.

McAloon, B. J. and Webster, B. C. (1969) *Theor. Chim. Acta* **15,** 385.

McDonald, W. J. and Thompson, J. C. (1966) *Phys. Rev.* **150,** 602.

Mackenzie, R. C. (1979) *J. Non-Cryst. Solids* **32,** 91.

MacKinnon, A. and Kramer, B. (1981) *Phys. Rev. Lett.* **47,** 1540.

Mackintosh, A. R. (1963) *J. Chem. Phys.* **38,** 1991.

MacMillan, W. L. (1981) *Phys. Rev.* **B24,** 2739.

McWhan, D. B. and Remeika, J. P. (1970) *Phys. Rev.* **B2,** 3734.

McWhan, D. B. and Rice, T. M. (1969) *Phys. Rev. Lett.* **22,** 887.

McWhan, D. B., Rice, T. M. and Schmidt, P. H. (1969) *Phys. Rev.* **177,** 1063.

McWhan, D. B., Remeika, J. P., Rice, T. M., Brinkman, W. F., Maita, J. P. and Menth, A. (1971) *Phys. Rev. Lett.* **27,** 941.

McWhan, D. B., Marezio, M., Remeika, J. P. and Dernier, P. D. (1972) *Phys. Rev.* **B5,** 2552.

McWhan, D. B., Remeika, J. P., Maita, J. P., Okinaka, H., Kosuge, K. and Kachi, S. (1973a) *Phys. Rev.* **B7,** 326.

McWhan, D. B., Menth, A., Remeika, J. P., Brinkman, W. F. and Rice, T. M. (1973b) *Phys. Rev.* **B7,** 1920.

McWhan, D. B., Remeika, J. P., Bader, S. D., Triplett, B. B. and Phillips, N. E. (1973c) *Phys. Rev.* **B7,** 3079.

Mahan, G. D. (1967a) *Phys. Rev. Lett.* **18,** 448.